Hoy D Orton

Orton and Sadler's Business Calculator and Accountant's Assistant

A cyclopdia of the most concise and practical methods of business calculations,

including many valuable labor-saving tables

Hoy D Orton

Orton and Sadler's Business Calculator and Accountant's Assistant
A cyclopdia of the most concise and practical methods of business calculations, including many valuable labor-saving tables

ISBN/EAN: 9783337312046

Printed in Europe, USA, Canada, Australia, Japan

Cover: Foto ©Suzi / pixelio.de

More available books at **www.hansebooks.com**

ORTON & SADLER'S

BUSINESS CALCULATOR AND ACCOUNTANTS ASSISTANT

A CYCLOPÆDIA
OF THE

Most Concise AND Practical Methods
OF
Business Calculations.

INCLUDING MANY VALUABLE LABOR-SAVING TABLES,

TOGETHER WITH

IMPROVED INTEREST TABLES,

DECIMAL SYSTEM:

SHOWING THE INTEREST ON FROM

$10 to $10,000—Rate, Ten per Cent. per Annum.

BY

HOY D. ORTON, AND W. H. SADLER,

Lightning Calculator, formerly teacher of Rapid Calculations at the U. S. Naval Academy.	President and Founder of Sadler's "Bryant & Stratton" Business College, Baltimore, Md.

*Designed for the practical use of the
Banker, Merchant, Accountant, Mechanic, Farmer,
Business Man and Student. Containing the shortest, simplest and
most rapid methods of Computing Numbers, adapted
to all kinds of business and every-day life.*

Written and arranged so as to be within the comprehension of every one
having the slightest knowledge of figures.

PRICE $1.00.

Sent to any part of the World on receipt of same.

BALTIMORE, MD,:
W. H. SADLER, Publisher.

1897.

Entered according to Act of Congress, in the year 1877, by

W. H. SADLER,

In the Office of the Librarian of Congress, at Washington.

The pages of reprint from "ORTON'S LIGHTNING CALCULATOR" are protected by copyrights of Hoy D. Orton, issued in 1866 and 1871.

N. B.—All rights reserved. Any infringement will be prosecuted to the fullest extent of the law.

AGENTS WANTED.

For particulars and territory, apply in person or address the publisher.

ORDERS.

Parties ordering the "CALCULATOR" should be particular to write plainly their Name, Residence, County and State.

Upon receipt of One Dollar a copy of the book will be forwarded, post-paid, to any address.

SADLER PUBLISHING CO.,

Nos. 10 & 12 N. CHARLES ST.,

BALTIMORE, MD.

PREFACE

THE principal features embodied in this work are simplicity and brevity.

There can be nothing new in principle, but so far as the authors' knowledge extends, their peculiar methods and abbreviations in the practical applications of the rules of Addition, Multiplication, Fractions, Percentage, Interest, Averaging Accounts, and Mensuration, have not heretofore been published, except such as are contained in the work of the senior author, known as "Orton's Lightning Calculator." As an endorsement of Professor Orton's original work on the subject of rapid or lightning calculations, it may be here stated that over 400,000 copies of that book have been sold. It is not the design of the authors of this work to make a text-book for the use of beginners in arithmetic, but to offer to those who have mastered the principles of addition, subtraction, multiplication and division, a guide to the practical application of that knowledge of arithmetic and calculations which is required daily in business and the affairs of life. There is no qualification more essential to success than facility in the rapid and accurate use of figures.

In view of the increasing demand for the original work, which has been several times revised, the present authors have decided to enlarge upon the subject. They present in this new volume—the result of their joint labors—the most extensive and comprehensive work of the kind ever offered to the public, in the full assurance that whoever will carefully study its pages will glean therefrom an abundant reward.

THE AUTHORS.

TABLE OF CONTENTS

PREFACE....	3
FRONTISPIECE—Illustration....	4
INTRODUCTION....	9
ADDITION....	12
" Lightning Method....	13
" " " —Table....	14
" Illustration....	15
" An Easy Way to Add .... 17 to	26
MULTIPLICATION—Illustrated....	26
" Short Methods.... 26 "	31
" Contractions.. .... 31 "	33
" " —Curious and Useful....	33
" Table of Squares....	34
FRACTIONS....	35
" Mental Operations....	36
" When the Sum of the Fractions is One.... 38 "	41
" When the Fractions have a like Denominator....	41
" Rapid Process of Multiplying Mixed Numbers.... 41 "	48
" When the Multiplier is an Aliquot Part of 100...	48
" Table of Aliquot Parts of 100 to 1,000....	48
" Counting-Room Exercises.... 49 "	53
" Illustration	53
" Division, with Analysis.... 53 "	55
" " by Boxing.... 55,	56
" Multiplication and Division.... 56 "	59
PERCENTAGE—As applied to Business	59
" Illustration	60
" Given Cost and Selling Price to find the Rate	62
" Given Profit and Rate to find the Cost....	63
" Given Amount and Rate to find the Cost....	63
" Given Proceeds, showing Loss and Rate, to find the Cost....	64
PROFIT AND LOSS—Illustration	65
" Short Business Methods....	66
" Table of Aliquot Parts	66
INTEREST—Showing Application of Percentage.... 67 "	72

5

CONTENTS.

DISCOUNT—Commercial	73
" True	74
" Bank	76
COMMISSION—Illustration	77
" Showing Application of Percentage	77 to 80
INSURANCE—Showing Application of Percentage	80
INVESTMENTS—Illustration	81
" Capital and Stocks, showing Application of Percentage	81 " 85
" Table for Investors	86
INTEREST Discount and Average—Illustration	87
" Simplified by Cancellation	93
" Short Practical Rules	100
BANKS AND BANKING—Illustration	104
INTEREST—Bankers' Method	105 " 114
" Lightning Method	114 " 117
" Merchants' Method	117 " 124
PARTIAL PAYMENTS—Notes, Bonds and Mortgages	124, 129
EQUATION OF PAYMENTS	129, 135
AVERAGING ACCOUNTS—Illustration	135
" Lightning Method	135 " 139
PARTNERSHIPS	139
" Settlements by Three Different Methods	140 " 144
GOLD TO CURRENCY—Gold at a Premium	144
CURRENCY TO GOLD— " " "	144, 145
MATURING NOTES, ETC.	145, 146
STERLING EXCHANGE—Illustration	147
" How Calculated	148 " 151
" Old Table	151
" New Method and Tables	152, 154
MARKING GOODS—Illustration	154
" Asking Price and Discounts	155, 159
" Rapid Process	159, 162
" Table for Marking all Goods Purchased by the Dozen	162
BASIS OF SUCCESS IN Business	163
LEDGER ACCOUNTS—Illustration, or the Science of Book-keeping Comprised in a Few Pages	164 " 175
HOW TO CLOSE THE LEDGER	175, 177
BALANCING BOOKS—Illustration	176
ERRORS IN TRIAL BALANCES—Illustration	178
" How to Detect Them	178 " 183
LUMBER MEASURING—Illustration	183
" Short, Practical Rules	183 " 186
MEASURING CORD WOOD—Illustration	186
" " Short, Practical Rules	186 " 189
ROUND TIMBER—Measuring—Illustration	189
" Short, Practical Rules	189 " 193
FLOORING— " " "	193
SQUARE TIMBER—Measuring—Illustration	194
" Short, Practical Rules	194 " 196
CISTERNS AND RESERVOIRS—Table of Capacities	196
" Illustration	196
" How to Measure their Contents	197 " 200

CONTENTS. 7

CASK GAUGING—Illustration	200
" Short, Practical Rules	203
MEASURING GRAIN—Illustration	203
" Size of Bins, How Ascertained	204
" Weights and Measures, U. S. Standard	204
WEIGHTS AND MEASURES ⎱ Table of Avoirdupois Weights, and	
BUSHELS TO POUNDS ⎰ No. of Pounds to the Bushel	205
IRON WEIGHTS—Used in Railroading—Table of Estimates	206
CORN IN CRIBS—Measuring—Illustration	207
" Practical Rules for Estimates....207 "	210
MEASURING HAY—Illustration.	
" Estimating Quantity in Stacks, Mows and Meadows....210 "	213
WEIGHT OF LIVE CATTLE—Illustration.	
" Weights Estimated by Measurement	214
BUILDERS' MEASUREMENTS—Illustration	215
" Bricklaying	216
" Tiling or Slating	217
" Walling	218
" Glazing	220
" Plumbing	221
" Masonry	222
" Plastering	223
SHORT RULES FOR THE MECHANIC—Illustration	225
SQUARE AND CUBE ROOTS....226 "	237
MENSURATION, OR PRACTICAL GEOMETRY....237 "	246
TABLE OF MULTIPLES	246

VALUABLE TABLES.

AVOIRDUPOIS WEIGHT—Illustration, with Tables	247
" Long or Iron Ton	247
" Iron and Lead	247
" Miscellaneous Table	248
APOTHECARIES' WEIGHT—Illustration, with Tables	248
" FLUID MEASURE	249
DRY MEASURE—Illustration and Tables	250
CUBIC OR SOLID MEASURE—Illustration, with Tables	251
MEASUREMENTS—Valuable Information Concerning	252
NAILS—Sizes and Number to the Pound	252
LIQUID MEASURE—Illustration, with Tables	253
MEASUREMENTS—LINEAR OR LONG—Illustration, with Tables	254
SURVEYORS' MEASURE	255
GEOGRAPHICAL AND ASTRONOMICAL CALCULATIONS—Table	255
SURFACE OR SQUARE MEASURE—Illustration, with Tables	256
SURVEYORS' SQUARE MEASURE, with Tables	257
CONTENTS OF FIELDS AND LOTS—Table	258
FENCING—Table showing the No. of Stakes, Rails and Posts Required in Fencing	258
TROY WEIGHT—Illustration, with Tables	259
DIAMOND WEIGHT	259
PAPER, BOOKS AND STATIONERY—Illustration, with various Tables	260

CONTENTS.

PRINTING—Type-setting	261
" Press-work	262
" Cost of Paper	263
" Table showing Cost of Paper by the Quire	264
BOOKS—Sizes and Styles	262
SHOEMAKERS' MEASURE	263
MEASUREMENT OF TIME—Illustration	265
" " Table	266
" " Circular Measure	266
LONGITUDE AND TIME—Table	267
TIME—How to Ascertain the Difference between Cities	267
TABLE—For Ascertaining the No. of Days between Two Dates	268
" Showing the Number of Days from any Day in one Month to the same Day in Another	269
ASTRONOMICAL CALCULATIONS	270 to 274
MONEY of the United States—Illustration	274
" France	275
" the German Empire	276
ARBITRATION OF EXCHANGE	277
VALUE OF FOREIGN COINS IN U. S. GOLD	278, 279
GOLD AND CURRENCY VALUES	280
BANK ACCOUNTS—Illustration—How to Transact Business with Banks	281 " 284
INTEREST—Commercial Rules	284, 285
INTEREST TABLES—Decimal System	286 " 290
" " Compound Interest	295
TIME REQUIRED FOR MONEY AT INTEREST TO DOUBLE	296
U. S. INTEREST RATES AND PENALTIES	297, 298
METRIC SYSTEM	299 to 305
TRADE DISCOUNTS	306
CLEARING HOUSE	308 " 318
WAGES—VALUE OF TIME	319
HOW TO OBTAIN WEALTH	320
VALUABLE TABLES	321 " 323

INTRODUCTION

QUANTITY is that which can be increased or diminished by augments or abatements of homogeneous parts. Quantities are of two essential kinds, *Geometrical* and *Physical*.

1. *Geometrical* quantities are those which occupy space; as *lines, surfaces, solids, liquids, gases*, etc.

2. *Physical* quantities are those which exist in the time, but occupy no space; they are known by their character and action upon geometrical quantities, as *attraction, light, heat, electricity* and *magnetism, colors, force, power*, etc.

To obtain the magnitude of a quantity we compare it with a part of the same; this part is imprinted in our mind as a *unit*, by which the whole is measured and conceived. No quantity can be measured by a quantity of another kind, but any quantity can be compared with any other quantity, and by such comparison arises what we call *calculation* or *Mathematics*.

INTRODUCTION.

Arithmetic means reckoning by numbers, **calculating**.

Notation means writing numbers.

Numeration means reading numbers.

Number is one or more things or **units**, as one, two, &c.

Unit or **one** is a single thing.

Numbers are represented by **figures**.

Figures are characters used in Arithmetic to represent numbers.

All numbers are represented by the ten following figures:

(*Written*)   0. 1. 2. 3. 4. 5. 6. 7. 8. 9.

          Cipher. one. two. three. four. five. six. seven. eight. nine.

(*Printed*)   0. 1. 2. 3. 4. 5. 6. 7. 8. 9.

These figures, except the cipher, are often called **Digits**.

Digit means the measure of a **finger's** breadth.

Figures were called **digits** from counting the fingers in reckoning.

    The character 0 is called a cipher, from the Arabic word *tsphara*, which signifies a *blank* or *void*. The uses of this character in numeration are so important, that its name, *cipher*, has been extended to the whole art of Arithmetic, which has been called to *cipher*, meaning *to work with figures*.

INTRODUCTION. 11

Standard Measures, to prevent error, are generally derived *from nature.* For example, measures of time, *from the time of the revolution of the earth about its axis;* of space, *from the length of a barley-corn,* taken from the middle of a full-grown ear; also, *from the circumference of the earth;* of weight, *from the weight of a grain of wheat,* taken as above; also, *from the weight of a definite quantity of distilled water;* of heat, *from the temperature of boiling water,* &c.

The four principal operations of Arithmetic are represented by the following signs:

+ *Plus* or more, the sign of Addition.
− *Minus* or less, " Subtraction.
× *Into* (multiplied by) " Multiplication.
÷ *By* (divided by) " Division.

When, in solving a question, only one operation is used, *the answer has a distinctive name.*

In addition, the answer is called the *sum*.

Subtraction,   "   "   { *Difference* or *Remainder.*
Multiplication, "   "   *Product.*
Division,   "   "   *Quotient.*

A sign made thus =, called *Equal to* or *Equals*, is placed between two quantities to show their equality; Thus, $1 + 1 = 2$ is read, *one plus one, equal to two;* or, more commonly and perhaps better, *one plus one equals two.*

ADDITION

To BE able to add two, three or four columns of figures at once, is deemed by many to be a Herculean task, and only to be accomplished by the gifted few, or, in other words, by mathematical prodigies. If we can succeed in dispelling this illusion, it will more than repay us; and we feel very confident that we can, if the student will lay aside all prejudice, bearing steadily in mind that to become proficient in any new branch or principle a little wholesome application is necessary. On the contrary, we can not teach a student who takes no interest in the matter, one who will always be a drone in society. Such men have no need of this principle.

If two, three, or more, columns can be carried up at a time, there must be some law or rule by which it is done. We have two principles of Addition; one for adding short columns, and one for adding very long columns. They are much alike, differing only in detail. When one is thoroughly learned, it is very easy to learn the second.

ADDITION TABLE.

The design of the table on the following page is to familiarize the student with the combination or grouping of figures so as to enable him instantly to see or read the result without stopping to add each figure separately.

In learning this table avoid spelling the figures, as 4 and 5 are 9, but take in the result 9 as soon as the eye catches the combination—do not consider the figures 4+5, but see them as 9. To illustrate: add 4+5+6+2, instead of saying 4 and 5 are 9 and 6 are 15 and 2 are 17, consider the combinations 4+5 as 9, 6+2 as 8; thus you really have but 9+8 to add instead of 4+5+6+2, producing a saving in time and mental work.

The science of

RAPID OR "LIGHTNING" ADDITION

Lies in the ability of the calculator to instantly see or take in the result of two or more figures regardless of their combination, without stopping to add each figure separately, *i. e.*, To read the result of figures as in reading a book, the pronunciation of a word is known, or the meaning of a sentence without the necessity of spelling or pronunciation of syllables.

After mastering this table the learner will be surprised at the rapidity he can add a column of figures, and he will soon find himself grouping or combining with ease and accuracy four and five figures at a time, instead of two as illustrated by the table.

TABLE OF ADDITION,

Showing the combination of the 9 significant figures, in groups of two only, and producing, when added together, results from 1 to 18.

Products.	Produced by combination or addition of the 9 significant figures.	Products.
1 =	1	= 1
2 =	1 1	= 2
3 =	2 1	= 3
4 =	3 2 1 2	= 4
5 =	4 3 1 2	= 5
6 =	5 4 3 1 2 3	= 6
7 =	6 5 4 1 2 3	= 7
8 =	7 6 5 4 1 2 3 4	= 8
9 =	8 7 6 5 1 2 3 4	= 9
10 =	9 8 7 6 5 1 2 3 4 5	= 10
11 =	9 8 7 6 2 3 4 5	= 11
12 =	9 8 7 6 3 4 5 6	= 12
13 =	9 8 7 4 5 6	= 13
14 =	9 8 7 5 6 7	= 14
15 =	9 8 6 7	= 15
16 =	9 8 7 8	= 16
17 =	9 8	= 17
18 =	9 9	= 18

ADDITION.

N. B. *The above process of addition is only recommended for beginners.*

Process.—For adding the above example, commence at the bottom of the right-hand column. Add thus: 12, 16, 22; then carry the 2 tens to the second column, then add thus, 8, 10, 18, 22, carry the two hundreds to the third column, and add the same way, 9, 13, 16, 23. Never permit yourself *for once* to add up a column in this manner, 3 and 9 are 12 and 4 are 16, and 6 are 22; it is just as easy to name the sum at *once, without naming the figures you add*, and three times as rapid.

ADDITION OF SHORT COLUMNS OF FIGURES.

ADDITION is the basis of all numerical operations, and is used in all departments of business To aid the business man in acquiring facility and accuracy in adding short columns of figures, the following method is presented as the best:

```
 274
 346
 134
 342
 727
 329
-----
2152
```

PROCESS.—Commence at the bottom of the right-hand column, add thus: 16, 22, 32; then carry the 3 tens to the second column; then add thus: 7, 14, 25; carry the 2 hundreds to the third column, and add the same way: 12, 16, 21. In this way you name the sum of two figures at once, which is quite as easy as it is to add one figure at a time. Never permit yourself *for once* to add up a column in this manner: 9 and 7 are 16, and 2 are 18 and 4 are 22, and 6 are 28, and 4 are 32. It is just as easy to name the result of two figures at once and four times as rapid.

The following method is recommended for the

ADDITION OF LONG COLUMNS OF FIGURES.

In the addition of long columns of figures which frequently occur in books of accounts, in order to add them with certainty, and, at the same time, with ease and expedition, study well the following method, which practice will render familiar, easy, rapid, and certain.

ADDITION. 17

THE EASY WAY TO ADD.

EXAMPLE 2—EXPLANATION.

Commence at 9 to add, and add as near 20 as possible, thus: $9+2+4+3=18$, place the 8 to the right of the 3, as in example; commence at 6 to add $6+4+8=18$; place the 8 to the right of the 8, as in example; commence at 6 to add $6+4+7=17$; place the 7 to the right of the 7, as in example; commence at 4 to add $4+9+3=16$; place the 6 to the right of the 3, as in example; commence at 6 to add $6+4+7=17$; place the 7 to the right of the 7, as in example; now, having arrived at the top of the column, we add the figures in the new column, thus: $7+6+7+8+8=36$; place the right hand figure of 36, which is a 6, under the original column, as in example, and add the left hand figure, which is a 3, to the number of figures in the new column; there are 5 figures in the new column, therefore $3+5=8$; prefix the 8 with the 6, under the original column, as in example; this makes 86, which is the sum of the column.

$$
\begin{array}{r}
7^7 \\
4 \\
6 \\
3^8 \\
9 \\
4 \\
7^7 \\
4 \\
6 \\
8^8 \\
4 \\
6 \\
3^8 \\
4 \\
2 \\
9 \\
\hline
86
\end{array}
$$

Remark 1.—If, upon arriving at the top of the column, there should be one, two or three figures whose sum will not equal 10, add them on to the sum of the figures of the new column, never placing an extra figure in the new column, unless it be an excess of units over ten.

Remark 2.—By this system of addition you can stop any place in the column, where the sum of the figures will equal 10 or the excess of 10; but the addition will be more rapid by your adding as near 20 as possible, because you will save the forming of extra figures in your new column.

EXAMPLE—EXPLANATION.

$2+6+7=15$, drop 10, place the 5 to the right of the 7; $6+5+4=15$, drop 10, place the 5 to the right of the 4, as in example; $8+3+7=18$, drop 10, place the 8 to the right of the 7, as in example; now we have an extra figure, which is 4; add this 4 to the top figure of the new column, and this sum on the balance of the figures in the new column, thus: $4+8+5+5=22$; place the right hand figure of 22 under the original column, as in example, and add the left hand figure of 22 to the number of figures in the new column, which are three, thus: $2+3=5$; prefix this 5 to the figure 2, under the original column; this makes 52. which is the sum of the column.

```
4
7⁵
3
8
4⁵
5
6
7⁵
6
2
—
52
```

ADDITION.

RULE.—*For adding two or more columns, commence at the right hand, or units' column; proceed in the same manner as in adding one column; after the sum of the first column is obtained, add all except the right hand figure of this sum to the second column, adding the second column the same way you added the first; proceed in like manner with all the columns, always adding to each successive column the sum of the column in the next lower order, minus the right hand figure.*

N. B. The small figures which we place to the right of the column when adding are called *integers*.

The addition by integers or by forming a new column, as explained in the preceding examples should be used only in adding very long columns of figures, say a long ledger column, where the footings of each column would be two or three hundred, in which case it is superior and much more easy than any other mode of addition; but in adding short columns it would be useless to form an extra column, where there is only, say, six or eight figures to be added. In making short additions, the following suggestions will, we trust, be of use to the accountant who seeks for information on this subject.

In the addition of several columns of figures, where they are only four or five deep, or when their respective sums will range from twenty-five to forty, the accountant should commence with the unit column, adding the sum of the first two figures to the sum of the next two, and so on, naming only the results, that is, the sum of every two figures.

In the present example in adding the unit column instead of saying 8 and 4 are 12 and 5 are 17 and 6 are 23, it is better to let the eye glide up the column reading only, 8, 12, 17, 23; and still better, instead of making a separate addition for each figure, group the figures thus: 12 and 11 are 23, and proceed in like manner with each column. For short columns this is a very expeditious way, and indeed to be preferred; but for long columns, the addition by integers is the most useful, as the mind is relieved at intervals and the mental labor of retaining the whole amount, as you add, is avoided, which is very important to any person whose mind is constantly employed in various commercial calculations.

  346
  235
  724
  598

In adding a long column, where the figures are of a medium size, that is, as many 8s and 9s as there are 2s and 3s, it is better to add about three figures at a time, because the eye will distinctly see that many at once, and the ingenious student will in a short time, if he adds by integers, be able to read the amount of three figures at a glance, or as quick, we might say, as he would read a single figure.

ADDITION. 21

Here we begin to add at the bottom of the unit column and add successively three figures at a time, and place their respective sums, minus 10, to the right of the last figure added; if the three figures do not make 10, add on more figures; if the three figures make 20 or more, only add two of the figures. The little figures that are placed to the right and left of the column are called integers. The integers in the present example, belonging to the units column, are 4, 4, 5, 4, 6, which we add together, making 23; place down 3 and add 2 to the number of integers, which gives 7, which we add to the tens and proceed as before.

$^5 26^6$
67
43
38^4
$^9 54$
62
87^5
$^5 65$
53
44^4
$^8 77$
33
84^4
$^3 56$
14
———
803

REASON.—In the above example, every time we placed down an integer we discarded a ten, and when we set down the 3 in the answer we discarded two tens; hence, we add 2 on to the number of integers to ascertain how many tens were discarded; there being 5 integers it made 7 tens, which we now add to the column of tens; on the same principle we might add between 20 and 30, always setting down a figure before we got to 30; then every integer set down would count for 2 tens, being discarded in the same way, it does in the present instance for one ten. When we add between 10 and 20, and in very long columns, it would be much better to go as near 30 as possible, and count 2 tens for every integer set down, in which case we would set down about one-half as many integers as when we write an integer for every ten we discard.

When adding long columns in a ledger or day-book, and where the accountant wishes to avoid the writing of extra figures in the book, he can place a strip of paper alongside of the column he wishes to add, and write the integers on the paper, and in this way the column can be added as convenient almost as if the integers were written in the book.

Perhaps, too, this would be as proper a time as any other to urge the importance of another good habit; I mean *that of making plain figures*. Some persons accustom themselves to making mere scrawls, and important blunders are often the result. If letters be badly made you may judge from such as are known; but if one figure be illegible, its value can not be inferred from the others. The vexation of the man who wrote for 2 or 3 monkeys, and had 203 sent him, was of far less importance than errors and disappointments sometimes result ing from this inexcusable practice.

We will now proceed to give some methods of proof. Many persons are fond of proving the correctness of work, and pupils are often instructed to do so, for the double purpose of giving them exercise in calculation and saving their teacher the trouble of reviewing their work.

There are special modes of proof of elementary operations, as by casting out threes or nines, or by changing the order of the operation, as in adding upward and then downward. In Addition, some prefer reviewing the work by performing the Addition downward, rather than repeating the ordinary operation. This is better, for if a mistake be inadvertently made in any calculation, and the same routine be again followed, we are very liable to fall again into the same error If, for instance, in running up a column of Addition you should say 84 and 8 are 93, you would be liable, in going over the same again, in the same way to slide insensibly into a similar error; but by beginning at a different point this is avoided.

This fact is one of the strongest objections to the plan of cutting off the upper line and adding it to the sum of the rest, and hence some cut off the lower line by which the spell is broken. The most thoughtless can not fail to see that adding a line *to* the sum of the rest, is the same as adding it in *with* the rest.

The mode of proof by casting out the nines and threes will be fully explained in a following chapter.

A very excellent mode of avoiding error in adding long columns is to set down the result of each column on some waste spot, observing to place the numbers successively a place further to the left each time, as in putting down the product figures in multiplication; and afterward add up the amount. In this way if the operator lose his count, he is not compelled to go back to units, but only to the foot of the column on which he is op-. erating. It is also true that the brisk accountant. who thinks on what he is doing, is less liable to err than the dilatory one who allows his mind to wander. Practice too will enable a person to read amounts without naming each figure, thus instead of saying 8 and 6 are 14, and 7 are 21 and 5 are 26. it is better to let the eye glide up the column, reading only 8, 14, 21, 26, etc.; and, still further, it is quite practicable to accustom one's self to group the figures in adding, and thus proceed very rapidly. Thus in adding the units' column, instead of adding a figure at a time, we see at a glance that 4 and 2 are 6, and that 5 and 3 are 8, then 6 and 8 are 14; we may then, if expert, add — constantly the sum of two or three figures at a time, and with practice this will be found highly advantageous in long columns of figures; or two or three columns may be added at a time, as the practiced eye will see that 24 and 62 are 86 almost as readily as that 4 and 2 are 6.

87
23
45
62
24

ADDITION. 25

Teachers will find the following mode of matching lines for beginners very convenient, as they can inspect them at a glance:

$$
\begin{array}{r}
\text{Add} \quad 7654384 \\
8786286 \\
3408698 \\
2345615 \\
1213713 \\
\hline
23408696
\end{array}
$$

In placing the above the lines are matched in pairs, the digits constantly making 9. In the above, the first and fourth, second and fifth are matched; and the middle is the *key line*, the result being just like it, except the units' place, which is as many less than the units in the key line as there are pairs of lines; and a similar number will occupy the extreme left. Though sometimes used as a puzzle, it is chiefly useful in teaching learners; and as the location of the key line may be changed in each successive example, if necessary, the artifice could not be detected. The number of lines is necessarily odd.

SHORT METHODS OF MULTIPLICATION.

RULE.—*Set down the smaller factor under the larger, units under units, tens under tens. Begin with the unit figure of the multiplier, multiply by it, first the units of the multiplicand, setting the units of the product, and reserving the tens to be added to the next product; now multiply the tens of the multiplicand by the unit figure of the multiplier, and the units of the multiplicand by tens figure of*

the multiplier; add these two products together, setting down the units of their sum, and reserving the tens to be added to the next product; now multiply the tens of the multiplicand by the tens figure of the multiplier, and set down the whole amount. This will be the complete product.

Remark.—Always add in the tens that are reserved as soon as you form the first product.

EXAMPLE 1.—EXPLANATION.

1. Multiply the units of the multiplicand by the unit figure of the multiplier, thus: 1×4 is 4; set the 4 down as in example.
2. Multiply the tens in the multiplicand by the unit figure in the multiplier, and the units in the multiplicand by the tens figure in the multiplier, thus: 1×2 is 2; 3×4 are 12, add these two products together, $2+12$ are 14, set the 4 down as in example, and reserve the 1 to be added to the next product. 3. Multiply the tens in the multiplicand by the tens figure in the multiplier, and add in the tens that were reserved, thus: 3×2 are 6, and $6+1=7$; now set down the whole amount, which is 7.

   24
   31
   ——
   744

EXAMPLE 2.—EXPLANATION.

1. Multiply units by units, thus: 4×3 are 12, set down the 2 and reserve the 1 to carry. 2. Multiply tens by units, and units by tens, and add in the one to carry on the

   53
   84
   ——
   4452

first product, then add these two products together, thus: 4×5 are 20+1 are 21, and 8×3 are 24, and 21+24 are 45, set down the 5 and reserve the 4 to carry to the next product. 3. Multiply tens by tens, and add in what was reserved to carry, thus: 8×5 are 40+4 are 44, now set down the whole amount, which is 44.

EXAMPLE 3.—EXPLANATION.

5×3 are 15, set down the 5 and carry the 1 to the next product; 5×4 are 20=1 are 21; 2×3 are 6, 21+6 are 27, set down the 7 and carry the 2; 2×4 are 8+2 are 10; now set down the whole amount.

```
  43
  25
 ———
1075
```

When the multiplicand is composed of three figures, and there are only two figures in the multiplier, we obtain the product by the following

RULE.—*Set down the smaller factor under the larger, units under units, tens under tens; now multiply the first upper figure by the unit figure of the multiplier, setting down the units of the product, and reserving the tens to be added to the next product; now multiply the second upper by units, and the first upper by tens, add these two products together, setting down the units figure of their sum, and reserving the tens to carry, as before; now multiply the third upper by units, and the second upper by tens, add these two products together, setting down the units figure of their sum, and reserving the tens to*

MULTIPLICATION.

carry, as usual; now multiply the third upper by tens, add in the reserved figure, if there is one, and set down the whole amount. This will be the complete product.

Remark.—One of the principal errors with the beginner, in this system of multiplication, is neglecting to add in the reserved figure. The student must bear in mind that the reserved figure is added on to the first product obtained after the setting down of a figure in the complete product.

EXAMPLE 1.—EXPLANATION.

Multiply first upper by units, 5×3 are 15, set down the 5, reserve the 1 to carry to the next product; now multiply second upper by units and first upper by tens, 5×2 are $10+1$ are 11, 4×3 are 12, add these products together; $11+12$ are 23, set down the 3, reserve the 2 to carry; now multiply third upper by units, and second upper by tens, add these two products together, always adding on the reserved figure to the first product; 5×1 are $5+2$ are 7, 4×2 are 8, and $7+8$ are 15, set down the 5, reserve the 1; now multiply third upper by tens, and set down the whole amount; 4×1 are $4+1$ are 5, set down the 5. This will give the complete product.

```
123
 45
----
5535
```

Multiply 32 by 45 in a single line.

Here we multiply 5×2 and set down and carry as usual; then to what you carry add 5×3 and 4× 2, which gives 24; set down 4 and carry 2 to 4×3, which gives 14 and completes the product.

  32
  45
 ——
 1440

Multiply 123 by 456 in a single line.

Here the first and second places are found as before; for the third, add 6×1, 5×2, 4×3, with the 2 you had to carry, making 30; set down 0 and carry 3; then drop the units' place and multiply the hundreds and tens crosswise, as you did the tens and units, and you find the thousand figure; then, dropping both units and tens, multiply the 4×1, adding the 1 you carried, and you have 5, which completes the product. The same principle may be extended to any number of places; but let each step be made perfectly familiar before advancing to another. Begin with two places, then take three, then four, but always practicing some time on each number, for any hesitation as you progress will confuse you.

  123
  456
 ———
 56088

N. B. The following mode of multiplying numbers will only apply where the sum of the two last or unit figures equal ten, and the other figures in both factors are the same.

CONTRACTIONS IN MULTIPLICATION.

To multiply when the unit figures added equal (10) and the tens are alike as 72 by 78, &c.

1st. Multiply the units and set down the result.

2d. Add 1 to either number in tens place and multiply by the other, and you have the complete product

EXAMPLE **PROCESS.**

Here because the sum of the units 4 and 6 are *ten* and the *tens* are alike; we simply say 4 times 6 are 24, and set down both figures of the product; then because 4 and 6 make *ten* we add 1 to 8, making 9, and 9 times 8 are 72, which completes the product.

```
  86
  84
 ----
7224
```

NOTE.—If the product of units do not contain ten the place of tens must be filled with a cipher

The above rule is useful in examples like the following:

2. What will 93 acres of land cost at 97 dollars per acre? Ans. $9021.

3. What will 89 pounds of tea cost at 81 cents per pound? Ans. $72.09.

In the above the product of 9 by 1 did not amount to ten, therefore 0 is placed in tens place.

4. Multiply 998 by 992. Ans. 990016.

In the above, because 2 and 8 are 10, we add 1 to 99, making 100; then 100 times 99 are 9900

EXAMPLE.

Multiply 79 by 71 in a single line.

Here we multiply 1×9 and set down the result, then we multiply the 7 in the multiplicand, increased by 1 by the 7 in the multiplier, 7×8, which gives 56 and completes the product.

```
   79
   71
 ----
 5609
```

EXAMPLE.

Multiply 197 by 193 in a single line.

Here we multiply 3×7 and set down the result, then we multiply the 19 in the multiplicand, increased by 1 by the 19 in the multiplier, 19×20, which gives 380 and completes the product.

```
   197
   193
 -----
 38021
```

EXAMPLE.

Multiply 996 by 994 in a single line.

Here we multiply 4×6 and set down the result, then we multiply the 99 in the multiplicand, increased by 1 by the 99 in the multiplier, 99×100, which gives 9900 and completes the product.

```
    996
    994
 ------
 990024
```

EXAMPLE.

Multiply 1208 by 1202 in a single line.

Here we multiply 2×8 and set down the result, then we multiply the 120 in the multiplicand, increased by 1 by the 120 in the multiplier, 120×121, which gives 14520 and completes the product.

```
    1208
    1202
 -------
 1452016
```

MULTIPLICATION.

CURIOUS AND USEFUL CONTRACTIONS.

To multiply any number, of two figures, by 11,
  RULE.— *Write the sum of the figures between them*
  1. Multiply 45 by 11.   Ans. 495
  Here 4 and 5 are 9, which write between 4 & 5
  2. Multiply 34 by 11.   Ans. 374.

  N. B. When the sum of the two figures is over 9, increase the left-hand figure by the 1 to carry.
  3. Multiply 87 by 11.   Ans. 957.

To square any number of 9s instantaneously, and without multiplying,
  RULE.— *Write down as many 9s less one as there are 9s in the given number, an 8, as many 0s as 9s, and a 1.*
  4. What is the square of 9999 ?   Ans. 99980001.
  EXPLANATION.—We have four 9s in the given number, so we write down three 9s, then an 8, then three 0s, and a 1.
  5. Square 999999.   Ans. 999998000001.

To square any number ending in 5,
  RULE.—*Omit the 5 and multiply the number, as it will then stand by the next higher number, and annex 25 to the product.*
  6. What is the square of 75 ?   Ans. 5625.
  EXPLANATION.—We simply say, 7 times 8 are 56, to which we annex 25.
  7. What is the square of 95 ?   Ans. 9025

TABLE OF SQUARES,
FROM 1 TO 104.

$1^2=1$	$27^2=729$	$53^2=2809$	$79^2=6241$
$2^2=4$	$28^2=784$	$54^2=2916$	$80^2=6400$
$3^2=9$	$29^2=841$	$55^2=3025$	$81^2=6561$
$4^2=16$	$30^2=900$	$56^2=3136$	$82^2=6724$
$5^2=25$	$31^2=961$	$57^2=3249$	$83^2=6889$
$6^2=36$	$32^2=1024$	$58^2=3364$	$84^2=7056$
$7^2=49$	$33^2=1089$	$59^2=3481$	$85^2=7225$
$8^2=64$	$34^2=1156$	$60^2=3600$	$86^2=7396$
$9^2=81$	$35^2=1225$	$61^2=3721$	$87^2=7569$
$10^2=100$	$36^2=1296$	$62^2=3844$	$88^2=7744$
$11^2=121$	$37^2=1369$	$63^2=3969$	$89^2=7921$
$12^2=144$	$38^2=1444$	$64^2=4096$	$90^2=8100$
$13^2=169$	$39^2=1521$	$65^2=4225$	$91^2=8281$
$14^2=196$	$40^2=1600$	$66^2=4356$	$92^2=8464$
$15^2=225$	$41^2=1681$	$67^2=4489$	$93^2=8649$
$16^2=256$	$42^2=1764$	$68^2=4624$	$94^2=8836$
$17^2=289$	$43^2=1869$	$69^2=4761$	$95^2=9025$
$18^2=324$	$44^2=1936$	$70^2=4900$	$96^2=9216$
$19^2=361$	$45^2=2025$	$71^2=5041$	$97^2=9409$
$20^2=400$	$46^2=2116$	$72^2=5184$	$98^2=9604$
$21^2=441$	$47^2=2209$	$73^2=5329$	$99^2=9801$
$22^2=484$	$48^2=2304$	$74^2=5476$	$100^2=10000$
$23^2=529$	$49^2=2401$	$75^2=5625$	$101^2=10201$
$24^2=576$	$50^2=2500$	$76^2=5776$	$102^2=10404$
$25^2=625$	$51^2=2601$	$77^2=5929$	$103^2=10609$
$26^2=676$	$52^2=2704$	$78^2=6084$	$104^2=10816$

NOTE.—To become familiar with the numbers shown in the above table, from 1 to 25, requires but little study and application upon the part of the pupil, and will prove of great benefit in mathematical calculations.

FRACTIONS.

Are one or more of the equal parts into which a unit or whole thing is divided.

All fractions express the division of units or things.

The fractional terms are, numerator and denominator.

The Numerator expresses the number of parts or units taken (it is therefore the dividend), and is written above the line.

The Denominator expresses the division of the equal parts or units (it is therefore the divisor), and is written below the line.

Fractions are written and read as follows:

$\frac{1}{2}$ or one-half, $\frac{1}{3}$ or one-third, $\frac{3}{4}$ or three-fourths.

The Quotient produced from dividing the numerator by the denominator of a fraction is its value.

$$\text{Thus}, - \frac{6}{2} = 3 - \frac{12}{4} = 3 - \frac{18}{2} = 9$$

The value of a fraction is less than 1 when the numerator is less than the denominator, and equals or exceeds 1 when the numerator equals or is greater than the denominator.

GENERAL PRINCIPLES GOVERNING FRACTIONS.

To increase or multiply a fraction,

Multiply the numerator or divide the denominator.

To decrease or divide a fraction,

Divide the numerator or multiply the denominator.

Multiplying or dividing both terms of a fraction by the same number *does not change its value.*

Fractions may be reduced, added, subtracted, multiplied, and divided.

Mental Operations in Fractions.

To square any number containing $\frac{1}{2}$, as $6\frac{1}{2}$, $9\frac{1}{2}$,

RULE.—*Multiply the whole number by the next higher whole number, and annex $\frac{1}{4}$ to the product.*

Ex. 1. What is the square of $7\frac{1}{2}$? Ans. $56\frac{1}{4}$.

We simply say, 7 times 8 are 56, to which we add $\frac{1}{4}$.

2. What will $9\frac{1}{2}$ lbs. beef cost at $9\frac{1}{2}$ cts. a lb.?
3. What will $12\frac{1}{2}$ yds. tape cost at $12\frac{1}{2}$ cts. a yd.?
4. What will $5\frac{1}{2}$ lbs. nails cost at $5\frac{1}{2}$ cts. a lb.?
5. What will $11\frac{1}{2}$ yds. tape cost at $11\frac{1}{2}$ cts. a yd.?
6. What will $19\frac{1}{2}$ bu. bran cost at $19\frac{1}{2}$ cts. a bu.?

REASON.—We multiply the whole number by the next higher whole number, because half of any number taken twice and added to its square is the same as to multiply the given number by ONE more than itself. The same principle will multiply any two *like* numbers together, when the sum of the fractions is ONE, as $8\frac{1}{8}$ by $8\frac{7}{8}$, or $11\frac{3}{8}$ by $11\frac{5}{8}$, etc. It is obvious that to multiply any number by any two fractions whose sum is ONE, that the sum of the products *must be the original number*, and adding the number to its square is simply to multiply it by ONE more than itself; for instance, to multiply $7\frac{1}{4}$ by $7\frac{3}{4}$, we simply say, 7 times 8 are 56, and then, to complete the multiplication, we add, of course, the product of the fractions ($\frac{3}{4}$ times $\frac{1}{4}$ are $\frac{3}{16}$), making $56\frac{3}{16}$ the answer.

MULTIPLICATION OF FRACTIONS.   37

Where the sum of the fractions is ONE.

To multiply any two *like* numbers together when the sum of the fractions is ONE,

RULE—*Multiply the whole number by the next higher whole number; after which, add the product of the fractions.*

N. B. In the following examples, the product of the fractions are obtained *first* for convenience.

PRACTICAL EXAMPLES FOR BUSINESS MEN.

Multiply $3\frac{1}{4}$ by $3\frac{3}{4}$ in a single line.

Here we multiply $\frac{1}{4} \times \frac{3}{4}$, which gives $\frac{3}{16}$, and set down the result; then we multiply the 3 in the multiplicand, increased by unity, by the 3 in the multiplier, 3×4, which gives 12 and completes the product.

$3\frac{3}{4}$
$3\frac{1}{4}$
———
$12\frac{3}{16}$

Multiply $7\frac{2}{5}$ by $7\frac{3}{5}$ in a single line.

Here we multiply $\frac{2}{5} \times \frac{3}{5}$, which gives $\frac{6}{25}$, and set down the result; then we multiply the 7 in the multiplicand, increased by unity, by the 7 in the multiplier, 7×8, which gives 56, and completes the product.

$7\frac{2}{5}$
$7\frac{3}{5}$
———
$56\frac{6}{25}$

Multiply $11\frac{1}{3}$ by $11\frac{2}{3}$ in a single line.

Here we multiply $\frac{2}{3} \times \frac{1}{3}$, which gives $\frac{2}{9}$, and set down the result; then we multiply the 11 in the multiplicand, increased by unity, by the 11 in the multiplier, 11×12, which gives 132 and completes the product.

$11\frac{1}{3}$
$11\frac{2}{3}$
———
$132\frac{2}{9}$

4

EXAMPLE.

Multiply $16\frac{2}{3}$ by $16\frac{1}{3}$ in a single line.

Here we multiply $\frac{1}{3}\times\frac{2}{3}$ which gives $\frac{2}{9}$, and set down the result, then we multiply the 16 in the multiplicand, increased by unity by the 16 in the multiplier, 16×17, which gives 272 and completes the product.

$$\begin{array}{r} 16\frac{2}{3} \\ 16\frac{1}{3} \\ \hline 272\frac{2}{9} \end{array}$$

EXAMPLE.

Multiply $29\frac{1}{2}$ by $29\frac{1}{2}$ in a single line.

Here we multiply $\frac{1}{2}\times\frac{1}{2}$ which gives $\frac{1}{4}$, and set down the result, then we multiply the 29 in the multiplicand, increased by unity by the 29 in the multiplier, 29×30, which gives 870 and completes the product.

$$\begin{array}{r} 29\frac{1}{2} \\ 29\frac{1}{2} \\ \hline 870\frac{1}{4} \end{array}$$

EXAMPLE.

Multiply $999\frac{3}{8}$ by $999\frac{5}{8}$ in a single line.

Here we multiply $\frac{5}{8}\times\frac{3}{8}$, which gives $\frac{15}{64}$, and set down the result, then we multiply the 999 in the multiplicand, increased by unity by the 999 in the multiplier, 999×1000, which gives 999000 and completes the product.

$$\begin{array}{r} 999\frac{3}{8} \\ 999\frac{5}{8} \\ \hline 999000\frac{15}{64} \end{array}$$

NOTE.—The system of multiplication introduced in the preceding examples, applies to all numbers. Where the sum of the fractions is *one*, and the whole numbers are alike, or differ by *one*, the learner is requested to study well these useful properties of numbers.

MULTIPLICATION OF FRACTIONS.

Where the sum of the Fractions is ONE.

To multiply any two numbers whose difference is *one*, and the sum of the fractions is *one*,

RULE.—*Multiply the larger number, increased by* ONE, *by the smaller number; then square the fraction of the larger number, and subtract its square from* ONE.

PRACTICAL EXAMPLES FOR BUSINESS MEN.

1. What will $9\frac{1}{4}$ lbs. sugar cost at $8\frac{3}{4}$ cts. a lb.?

Here we multiply 9, increased by 1, by 8, thus, 8×10 are 80, and set down the result: then from 1 we subtract the square of $\frac{1}{4}$, thus, $\frac{1}{4}$ squared is $\frac{1}{16}$, and 1 less $\frac{1}{16}$ is $\frac{15}{16}$.

$9\frac{1}{4}$
$8\frac{3}{4}$
——
$80\frac{15}{16}$

2. What will $8\frac{2}{3}$ bu. coal cost at $7\frac{1}{3}$ cts. a bu.?

Here we multiply 8, increased by 1, by 7, thus, 7 times 9 are 63, and set down the result; then from 1 we subtract the square of $\frac{1}{3}$, thus, $\frac{1}{3}$ squared is $\frac{1}{9}$, and 1, less $\frac{1}{9}$, is $\frac{8}{9}$.

$8\frac{2}{3}$
$7\frac{1}{3}$
——
$63\frac{8}{9}$

3. What will $11\frac{2}{13}$ bu. seed cost at $\$10\frac{11}{13}$ a bu.?

Here we multiply 11, increased by 1, by 10, thus, 10 times 12 are 120, and set down the result; then from 1 we subtract the square of $\frac{2}{13}$, thus, $\frac{2}{13}$ squared is $\frac{4}{169}$, and 1 less $\frac{4}{169}$ is $\frac{165}{169}$.

$11\frac{2}{13}$
$10\frac{11}{13}$
——
$120\frac{165}{169}$

4. How many square inches in a floor $99\frac{3}{4}$ in. wide and $98\frac{1}{4}$ in. long? Ans. $9800\frac{15}{16}$.

METHOD OF OPERATION.

EXAMPLE.

Multiply $6\frac{1}{4}$ by $6\frac{1}{4}$ in a single line.

Here we add $6\frac{1}{4}+\frac{1}{4}$, which gives $6\frac{1}{2}$; this multiplied by the 6 in the multiplier, $6\times 6\frac{1}{2}$, gives 39, to which we add the product of the fractions, thus $\frac{1}{4}\times\frac{1}{4}$ gives $\frac{1}{16}$, added $39\frac{1}{16}$ to 39 completes the product.

$6\frac{1}{4}$
$6\frac{1}{4}$
———

EXAMPLE.

Multiply $11\frac{1}{4}$ by $11\frac{3}{4}$ in a single line.

Here we would add $11\frac{1}{4}+\frac{3}{4}$, which gives 12; this multiplied by the 11 in the multiplier gives 132, to which we add the product of the fractions, thus $\frac{3}{4}\times\frac{1}{4}$ gives $\frac{3}{16}$, which $132\frac{3}{16}$ added to 132 completes the product.

$11\frac{1}{4}$
$11\frac{3}{4}$
———

EXAMPLE.

Multiply $12\frac{1}{2}$ by $12\frac{3}{4}$ in a single line.

Here we add $12\frac{1}{2}+\frac{3}{4}$, which gives $13\frac{1}{4}$; this multiplied by the 12 in the multiplier, $12\times 13\frac{1}{4}$, gives 159, to which add the product of the fractions, thus $\frac{3}{4}\times\frac{1}{2}$ gives $\frac{3}{8}$, which added to 159 completes the product.

$12\frac{1}{2}$
$12\frac{3}{4}$
———
$159\frac{3}{8}$

MULTIPLICATION OF FRACTIONS.

Where the Fractions have a Like Denominator.

To multiply any two *like* numbers together, each of which has a fraction with a *like* denominator, as $4\frac{3}{8}$ by $4\frac{7}{8}$, or $11\frac{1}{4}$ by $11\frac{3}{4}$, or $10\frac{2}{3}$ by $10\frac{1}{3}$, etc.

RULE.—*Add to the multiplicand the fraction of the multiplier, and multiply this sum by the whole number; after which, add the product of the fractions.*

PRACTICAL EXAMPLES FOR BUSINESS MEN.

N. B. In the following example, the sum of the fractions is ONE.

1. What will $9\frac{3}{4}$ lbs. beef cost at $9\frac{1}{4}$ cts. a lb.?

The sum of $9\frac{3}{4}$ and $\frac{1}{4}$ is ten, so we simply say, 9 times 10 are 90; then we add the product of the fractions, $\frac{1}{4}$ times $\frac{3}{4}$ are $\frac{3}{16}$. $9\frac{3}{4}$
$9\frac{1}{4}$
$90\frac{3}{16}$

N. B. In the following example, the sum of the fractions is *less* than ONE.

2. What will $8\frac{1}{4}$ yds. tape cost at $8\frac{2}{4}$ cts. a yd.?

The sum of $8\frac{1}{4}$ and $\frac{2}{4}$ is $8\frac{3}{4}$, so we simply say, 8 times $8\frac{3}{4}$ are 70; then we add the product of the fractions, $\frac{2}{4}$ times $\frac{1}{4}$ are $\frac{2}{16}$ or $\frac{1}{8}$. $8\frac{1}{4}$
$8\frac{2}{4}$
$70\frac{1}{8}$

N. B. In the following example, the sum of the fractions is *greater* than ONE.

3. What will $4\frac{3}{8}$ yds. cloth cost at \$$4\frac{7}{8}$ a yd.?

The sum of $4\frac{3}{8}$ and $\frac{7}{8}$ is $5\frac{1}{4}$, so we simply say, 4 times $5\frac{1}{4}$ are 21; then we add the product of the fractions, $\frac{7}{8}$ times $\frac{3}{8}$ are $\frac{21}{64}$. $4\frac{3}{8}$
$4\frac{7}{8}$
$21\frac{21}{64}$

N. B. Where the fractions have different denominators reduce them to a common denominator.

Rapid Process of Multiplying Mixed Numbers.

A valuable and useful rule for the accountant in the practical calculations of the counting-room.

To multiply any two numbers together, each of which involves the fraction $\frac{1}{2}$, as $7\frac{1}{2}$ by $9\frac{1}{2}$, etc.,

RULE.—*To the product of the whole numbers add half their sum plus $\frac{1}{4}$.*

EXAMPLES FOR MENTAL OPERATIONS.

1. What will $3\frac{1}{2}$ doz. eggs cost at $7\frac{1}{2}$ cts. a doz.?

Here the sum of 7 and 3 is 10, and half this sum is 5, so we simply say, 7 times 3 are 21 and 5 are 26, to which we add $\frac{1}{4}$.

$$\begin{array}{r} 3\frac{1}{2} \\ 7\frac{1}{2} \\ \hline 26\frac{1}{4} \end{array}$$

N. B. If the sum be an odd number, call it one less to make it even, and in such cases the fraction must be $\frac{3}{4}$.

2. What will $11\frac{1}{2}$ lbs. cheese cost at $9\frac{1}{2}$ cts. a lb.?
3. What will $8\frac{1}{2}$ yds. tape cost at $15\frac{1}{2}$ cts. a yd.?
4. What will $7\frac{1}{2}$ lbs. rice cost at $13\frac{1}{2}$ cts. a lb.?
5. What will $10\frac{1}{2}$ bu. coal cost at $12\frac{1}{2}$ cts. a bu.?

REASON.—In explaining the above rule, we add half their sum because half of either number added to half the other would be half their sum, and we add $\frac{1}{4}$ because $\frac{1}{2}$ by $\frac{1}{2}$ is $\frac{1}{4}$. The same principle will multiply any two numbers together, each of which has the same fraction; for instance, if the fraction was $\frac{1}{3}$, we would add one-third their sum; if $\frac{3}{4}$, we would add three-fourths their sum, etc.; and then, to complete the multiplication, we would add, of course, the product of the fractions.

MULTIPLICATION OF FRACTIONS.

GENERAL RULE.

For multiplying any two numbers together, each of which involves the same fraction.

To the product of the whole numbers, add the product of their sum by either fraction; after which, add the product of their fractions.

EXAMPLES FOR MENTAL OPERATIONS.

1. What will $11\frac{3}{4}$ lbs. rice cost at $9\frac{3}{4}$ cts. a lb.?

Here the sum of 9 and 11 is 20, and three-fourths of this sum is 15, so we simply say, 9 times 11 are 99 and 15 are 114, to which we add the product of the fractions ($\frac{9}{16}$).

$$\begin{array}{r} 11\frac{3}{4} \\ 9\frac{3}{4} \\ \hline 114\frac{9}{16} \end{array}$$

2. What will $7\frac{2}{3}$ doz. eggs cost at $8\frac{2}{3}$ cts. a doz.?
3. What will $6\frac{3}{4}$ bu. coal cost at $6\frac{3}{4}$ cts. a bu.?
4. What will $45\frac{2}{3}$ bu. seed cost at $3\frac{2}{3}$ dol. a bu.?
5. What will $3\frac{5}{8}$ yds. cloth cost at $5\frac{5}{8}$ dol. a yd.?
6. What will $17\frac{3}{4}$ ft. boards cost at $13\frac{2}{3}$ cts a ft.?
7. What will $18\frac{3}{4}$ lbs. butter cost at $18\frac{3}{4}$ cts. a lb.?

N. B. If the product of the sum by either fraction is a whole number with a fraction, it is better to reserve the fraction until we are through with the whole numbers, and then add it to the product of the fractions; for instance, to multiply $3\frac{1}{4}$ by $7\frac{1}{4}$, we find the sum of 7 and 3, which is 10, and one-fourth of this sum is $2\frac{1}{2}$; setting the $\frac{1}{2}$ down in some waste spot, we simply say, 7 times 3 are 21 and 2 are 23; then, adding the $\frac{1}{2}$ to the product of the fractions ($\frac{1}{16}$), gives $\frac{9}{16}$, making $23\frac{9}{16}$, Ans.

Rapid Process of Multiplying all Mixed Numbers.

N. B. Let the student remember that this is a general and universal rule.

GENELAL RULE.

To multiply any two mixed numbers **together**,

1st. *Multiply the whole numbers together.*
2d. *Multiply the upper digit by the lower fraction.*
3d. *Multiply the lower digit by the upper fraction.*
4th. *Multiply the fractions together.*
5th. *Add these* FOUR *products together.*

N. B. This rule is so simple, so useful, and so true that every banker, broker, merchant, and clerk should post it up for reference and use.

PRACTICAL EXAMPLES FOR BUSINESS MEN

N. B. The following method is recommended to beginners:

EXAMPLE.—Multiply $12\frac{3}{4}$ by $9\frac{3}{4}$.

$$\begin{array}{r} 12\frac{3}{4} \\ 9\frac{3}{4} \\ \hline 108 \\ 9 \\ 6 \\ 0\frac{6}{12} \\ \hline 123\frac{6}{12} \end{array}$$

1st. We multiply the whole numbers.
2d. Multiply 12 by $\frac{3}{4}$ and write it down.
3d. Multiply 9 by $\frac{3}{4}$ and write it down.
4th. Multiply $\frac{3}{4}$ by $\frac{3}{4}$ and write it down.
5th. Add these *four* products together, and we have the complete result.

N. B. When the student has become familiar with the above process, it is better to do the intermediate work in the head, and, instead of setting down the partial products, add them in the mind as you pass along, and thus proceed very rapidly.

MULTIPLICATION OF MIXED NUMBERS.

Multiply $8\frac{1}{5}$ by $10\frac{1}{4}$.

Here we simply say 10 times 8 are 80 and $\frac{1}{4}$ of 8 is 2, making 82, and $\frac{1}{5}$ of 10 is 2, which makes 84; then $\frac{1}{4}$ times $\frac{1}{5}$ is $\frac{1}{20}$, making $84\frac{1}{20}$ the answer.

$8\frac{1}{5}$
$10\frac{1}{4}$
———
$84\frac{1}{20}$

PRACTICAL BUSINESS METHOD

For Multiplying all Mixed Numbers.

Merchants, grocers, and business men generally, in multiplying the mixed numbers that arise in the practical calculations of their business, only care about having the answer correct to the nearest cent; that is, they disregard the fraction. When it is a half cent or more, they call it another cent; if less than half a cent, they drop it. And the object of the following rule is to show the business man the easiest and most rapid process of finding the product to the nearest unit of any two numbers, one or both of which involves a fraction.

GENERAL RULE.

To multiply any two numbers to the nearest unit,

1st. *Multiply the whole number in the multiplicand by the fraction in the multiplier to the nearest unit.*

2d. *Multiply the whole number in the multiplier by the fraction in the multiplicand to the nearest unit*

3d. *Multiply the whole numbers together and add the three products in your mind as you proceed.*

N. B. In actual business the work can generally be done mentally for only easy fractions occur in *business*.

N. B. This rule is so simple and so true, according to all business usage, that every accountant should make himself perfectly familiar with its application. There being no such thing as a fraction to add in, there is scarcely any liability to error or mistake. By no other arithmetical process can the result be obtained by so few figures.

EXAMPLES FOR MENTAL OPERATION.

EXAMPLE FIRST.

Multiply $11\frac{1}{4}$ by $8\frac{1}{4}$ by business method. $11\frac{1}{4}$

Here $\frac{1}{4}$ of 11 to the nearest unit is 3, and $\frac{1}{4}$ of $8\frac{1}{4}$
8 to the nearest unit is 3, making 6, so we simply say, 8 times 11 are 88 and 6 are 94, Ans.   94

REASON.—$\frac{1}{4}$ of 11 is nearer 3 than 2, and $\frac{1}{4}$ of 8 is nearer 3 than 2. Make the nearest whole number the quotient.

EXAMPLE SECOND.

Multiply $7\frac{3}{4}$ by $9\frac{2}{8}$ by business method.

Here $\frac{3}{4}$ of 7 to the nearest unit is 3, and $\frac{1}{4}$ $7\frac{3}{4}$
of 9 to the nearest unit is 7; then 3 plus 7 $9\frac{2}{8}$
is 10, so we simply say, 9 times 7 are 63 and
10 are 73, Ans.   73

EXAMPLE THIRD.

Multiply $23\frac{1}{4}$ by $19\frac{1}{4}$ by business method.

Here $\frac{1}{4}$ of 23 to the nearest unit is 6, and $23\frac{1}{4}$
$\frac{1}{3}$ of 19 to the nearest unit is 6; then 6 plus $19\frac{1}{4}$
6 is 12, so we simply say, 19 times 23 are
437 and 12 are 449, Ans.   449

N. B. In multiplying the whole numbers together, always use the single-line method.

MULTIPLICATION OF MIXED NUMBERS. 47

EXAMPLE FOURTH.

Multiply 128¾ by 25 by business method.

Here ¾ of 25 to the nearest unit is 17, so   128¾
we simply say, 25 times 128 are 3200 and   25
17 are 3217, the answer.   3217

PRACTICAL EXAMPLES FOR BUSINESS MEN.

1. What is the cost of 17½ lbs. sugar at 18¾ cts. per lb.?

Here ¾ of 17 to the nearest unit is 13,   17½
and ½ of 18; is 9 13 plus 9 is 22, so we   18¾
simply say, 18 times 17 are 306 and 22 are   ———
328, the answer. $3.28

2. What is the cost of 11 lbs. 5 oz. of butter at 33⅓ cts. per lb.?

Here ⅓ of 11 to the nearest unit is 4,   11 5⁄16
and 5⁄16 of 33 to the nearest unit is 10;   33⅓
then 4 plus 10 is 14, so we simply say, 33   ———
times 11 are 363, and 14 are 377, Ans. $3.77

3. What is the cost of 17 doz. and 9 eggs at 12½ cts. per doz.?

Here ½ of 17 to the nearest unit is 9,   17 9⁄12
and 9⁄12 of 12 is 9; then nine plus 9 is 18,   12½
so we simply say, 12 times 17 are 204 and   ———
18 are 222, the answer. $2.22

4. What will be the cost of 15¾ yds. calico at 12½ cts. per yd.?   Ans. $1.97.

Where the Multiplier is an Aliquot Part of 100.

Merchants in selling goods generally make the price of an article some aliquot part of 100, as in selling sugar at 12½ cents a pound or 8 pounds for 1 dollar, or in selling calico for 16⅔ cents a yard or 6 yards for 1 dollar, etc. And to become familiar with all the aliquot parts of 100, so that you can apply them readily when occasion requires, is perhaps the most useful, and, at the same time, one of the easiest arrived at of all the computations the accountant must perform in the practical calculations of the counting-room.

TABLE OF THE ALIQUOT PARTS OF 100 AND 1000

N. B. Most of these are used in business.

12½ is $\frac{1}{8}$ part of 100.	8⅓ is $\frac{1}{12}$ part of 100.
25 is $\frac{2}{8}$ or $\frac{1}{4}$ of 100.	16⅔ is $\frac{2}{12}$ or $\frac{1}{6}$ of 100.
37½ is $\frac{3}{8}$ part of 100.	33⅓ is $\frac{4}{12}$ or $\frac{1}{3}$ of 100.
50 is $\frac{4}{8}$ or $\frac{1}{2}$ of 100.	66⅔ is $\frac{8}{12}$ or $\frac{2}{3}$ of 100.
62½ is $\frac{5}{8}$ part of 100.	83⅓ is $\frac{10}{12}$ or $\frac{5}{6}$ of 100.
75 is $\frac{6}{8}$ or $\frac{3}{4}$ of 100.	125 is $\frac{1}{8}$ part of 1000.
87½ is $\frac{7}{8}$ part of 100.	250 is $\frac{2}{8}$ or $\frac{1}{4}$ of 1000.
6¼ is $\frac{1}{16}$ part of 100.	375 is $\frac{3}{8}$ part of 1000.
18¾ is $\frac{3}{16}$ part of 100.	625 is $\frac{5}{8}$ part of 1000.
31¼ is $\frac{5}{16}$ part of 100.	875 is $\frac{7}{8}$ part of 1000.

To multiply by an aliquot part of 100,

RULE.—*Add two ciphers to the multiplicand, then take such part of it as the multiplier is part of* 100.

N. B. If the multiplicand is a mixed number reduce the fraction to a decimal of two places before dividing.

COUNTING-ROOM EXERCISES.

EXAMPLES.—1. Multiply 424 by 25.

As $25 = \frac{1}{4}$ of 100, divide 42400 by $4 = 10600$.

N. B. If the multiplicand is a mixed number, reduce the fraction to a decimal of two places before dividing.

2. Give the cost of $12\frac{1}{2}$ yds. cloth @ $18\frac{3}{4}$c. per yd.

Process.—$12\frac{1}{2} = \frac{1}{8}$; changing $18\frac{3}{4}$ to a decimal, we have $18.75 \div 8 = \$2.34\frac{3}{8}$.

Note.—Aliquot parts may be conveniently used when the multiplier is but little more or less than an aliquot part.

3. Multiply 24 by $17\frac{2}{3}$.

1st. Multiply 24 by $16\frac{2}{3}$ (the *one-sixth* of 100).

Thus $24 \times 16\frac{2}{3} = 2400 \div 6 = 400$

As $17\frac{2}{3} = 16\frac{2}{3} + 1$ multiply 24 by $1 = 24$

Hence $24 \times 17\frac{2}{3} =$ the two products. $\overline{424}$

3. To multiply any number by 125 add three ciphers, and divide by 8.

Multiply 3467 by 125.   Product, 433375.

$$8)\overline{3467000}$$
$$433375$$

NOTE.—By annexing three ciphers the number is increased one thousand times; and by dividing by 8, the quotient will be only one-eighth of 1000, that is 125 times.

4. To multiply any number by $16\tfrac{2}{3}$ add two ciphers, and divide by 6.

Multiply 3768 by $16\tfrac{2}{3}$.   Product, 62800.

$$6)\overline{376800}$$
$$62800$$

5. To multiply any number by $166\tfrac{2}{3}$ add three ciphers, and divide by 6.

Multiply 7875 by $166\tfrac{2}{3}$.   Product, 1312500.

$$6)\overline{7875000}$$
$$1312500$$

6. To multiply any number by $33\tfrac{1}{3}$ add two ciphers, and divide by 3.

Multiply 9879 by $33\tfrac{1}{3}$.   Product, 329300.

$$3)\overline{987900}$$
$$329300$$

COUNTING-ROOM EXERCISES.   51

RATIONALE.—As in the last case, by annexing two ciphers, we increase the multiplicand one hundred times; and by dividing the number by 3, we only increase the multiplicand thirty-three and one-third times, because $33\frac{1}{3}$ is one-third of 100.

4. To multiply any number by $333\frac{1}{3}$ add three ciphers, and divide by 3.

Multiply 4797 by $333\frac{1}{3}$.   Product, 1599000.

$$3)\overline{4797000}$$
$$1599000$$

5. To multiply any number by $6\frac{2}{3}$ add two ciphers, and divide by 15; or add one cipher and multiply by $\frac{2}{3}$.

Multiply 1566 by $6\frac{2}{3}$.

$$15)\overline{156600}$$
$$10440 \text{ First method.}$$

$$15660$$
$$2$$
$$3)\overline{31320}$$
$$10440 \text{ Second method.}$$

6. To multiply any number by $66\frac{2}{3}$ add three ciphers, and divide by 15; or add two ciphers and multiply by $\frac{2}{3}$.

Multiply 3663 by 66⅔.

```
    15)3663000
       ———————
        244200  First method.
       366300
            2
       ———————
     3)732600
       ———————
        244200  Second method.
```

7. To multiply any number by 8⅓ add two ciphers, and divide by 12.

Multiply 2889 by 8⅓. Product, 24075.

```
       12)288900
          ——————
            24075
```

8. To multiply any number by 83⅓ add three ciphers, and divide by 12.

Multiply 7695 by 83⅓ Product, 641250.

```
       12)7695000
          ———————
            641250
```

9. To multiply any number by 6¼ add two ciphers, and divide by 16 or its factors—4×4.

Multiply 7696 by 6¼. Product, 48100.

```
        4)769600
          ——————
        4)192400
          ——————
            48100
```

DIVISION OF FRACTIONS,
WITH ANALYSIS.

As the base of all numbers, whether whole or fractional, are of the same value, inverting any number simply demonstrates the number of times it is contained in a single unit.

As the unit takes the place of the number, so must the number *take the place of the unit.*

EXAMPLE 1.—Divide 6 by 7. Ans. $\frac{6}{7}$.

By inverting the divisor we find $\frac{7}{1}$, $\frac{1}{7}$; now if 7 is contained in one unit $\frac{1}{7}$ of one time, it is contained in 6, six times $\frac{1}{7}$, or $\frac{6}{7}$ *Ans.*

EXAMPLE 2.—Divide 5 by $\frac{4}{5}$. Ans. $6\frac{1}{4}$.

$\frac{4}{5}$ is contained in a unit $\frac{5}{4}$ times, therefore it is contained in $\frac{5}{1}$; $\frac{5}{1} \times \frac{5}{4} = \frac{25}{4}$, or $6\frac{1}{4}$ *Ans.*

53

EXAMPLE 3.—Divide $8\frac{1}{2}$ by $\frac{3}{4}$.   Ans. $11\frac{1}{3}$.
By inverting $\frac{3}{4}$ we have $\frac{4}{3}$, the number of times it is contained in a unit; therefore it is contained $8\frac{1}{2}$, $8\frac{1}{2}$ times $\frac{4}{3}$ or $\frac{17}{2} \times \frac{4}{3} = \frac{68}{6}$ or $11\frac{2}{6} = 11\frac{1}{3}$ Ans. (Note. $8\frac{1}{2} \times \frac{4}{3} = \frac{34}{3}$ or $11\frac{1}{3}$.)

EXAMPLE 4.—Divide $6\frac{1}{2}$ by 3.   Ans. $2\frac{1}{6}$.
As 3 is contained in a unit $\frac{1}{3}$ of one time, $6\frac{1}{2}$ is contained $6\frac{1}{2}$ times $\frac{1}{3}$, or thus,

$$\tfrac{13}{2} \div 3 = \tfrac{13}{2} \times \tfrac{1}{3} = \tfrac{13}{6} \text{ or } 2\tfrac{1}{6} \text{ Ans.}$$

RULE.—*To ascertain the number of times the divisor is contained in a unit, invert the divisor and multiply by the units in the dividend.*

EXAMPLE 5.—Divide 3 by $\frac{4}{7}$.   Ans. $5\frac{1}{4}$.
Solution.—$\frac{4}{7}$ is contained in a unit $\frac{7}{4}$ times, therefore it is contained in 3, 3 times $\frac{7}{4}$ or $\frac{21}{4} = 5\frac{1}{4}$ Ans.

NOTE.—Inverting the divisor shows how often it is contained in a unit, and multiplying this number by the dividend gives the quotient of all examples in

DIVISION OF FRACTIONS.

EXAMPLE 6.—Divide $\frac{1}{5}$ by $\frac{1}{4}$.   Ans. $\frac{4}{5}$.
Solution.—$\frac{1}{4}$ is contained in a unit 4 times, therefore it is contained in $\frac{1}{5}$, $\frac{1}{5}$ of 4 times, or $\frac{4}{5}$ times Ans.

EXAMPLE 7.—Divide $\frac{3}{4}$ of $\frac{2}{5}$ by $\frac{4}{5}$ of $\frac{1}{2}$.

Statement. $\dfrac{\frac{3}{4} \text{ of } \frac{2}{5}}{\frac{4}{5} \text{ of } \frac{1}{2}} = \frac{3}{4}$

NOTE.—Simply reject or cancel factors in the dividend and divisor, *thus you have $\frac{3}{4}$ as the quotient.*

DIVISION OF FRACTIONS.

EXAMPLE 8.—Divide $\frac{3}{4}$ of $\frac{2}{3}$ of $\frac{2}{5}$ by $\frac{3}{10}$ of $\frac{4}{9}$ of $\frac{2}{5}$.

1st Statement.—$\left.\begin{array}{c}\frac{3}{4}\text{ of }\frac{2}{3}\text{ of }\frac{2}{5}\\ \frac{3}{10}\text{ of }\frac{4}{9}\text{ of }\frac{2}{5}\end{array}\right\}$ Rejecting common factors and we have for the result $\frac{25}{12}$ or $2\frac{1}{12}$.

2d Statement.—

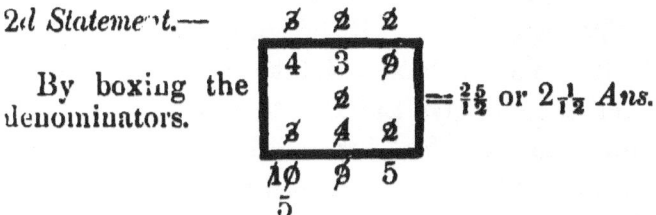

By boxing the denominators. $=\frac{25}{12}$ or $2\frac{1}{12}$ *Ans.*

RULE.—*Reject or cancel factors common to the divisor and dividend. Multiply the remaining terms between lines* (or inside the box) *together for the desired denominator; and the terms above and below the lines* (or outside the box) *for the desired numerator.*

NOTE.—The above rule is commended as the most simple one devised for the division of fractions.

EXAMPLE 9.—Divide $\frac{3}{4}$ of $\frac{2}{5}$ by $\frac{4}{5}$ of $\frac{1}{2}$.

The application of the above rule to Example 9 produces the following statement, so simple that it can readily be understood by the most ordinary pupil.

Draw a figure of four sides, representing a box, then write the example as follows:

$$\begin{array}{cc} 3 & 2 \\ \boxed{\begin{array}{cc} 4 & 5 \\ 4 & 1 \end{array}} \\ 5 & 2 \end{array} = \frac{60}{80} \text{ or } \frac{3}{4}.$$

The numerator of the fraction will be found outside the box, which, multiplied together, will give the desired numerator, thus,

$3 \times 2 \times 5 \times 2 = 60$

The denominator will be found upon the inside of the box, which, multiplied together, will give the desired denominator, thus, $4 \times 5 \times 4 \times 1 = 80$

} $\frac{60}{80}$ or $\frac{3}{4}$.

NOTE. When factors are common to each other always take advantage of cancellation.

An investigation of our new method of treating the division of fractions will prove, no doubt, to be the most simple and practical ever yet devised.

MULITIPLICATION AND DIVISION.

To multiply $\frac{1}{2}$, is to take the multiplicand $\frac{1}{2}$ of *one time;* that is, take $\frac{1}{2}$ of it, or divide it by 2.

To multiply by $\frac{1}{3}$, take a third of the multiplicand, that is, divide it by 3.

To multiply by $\frac{2}{3}$, take $\frac{1}{3}$ first, and multiply that by 2; or, multiply by 2 first, and divide the product by 3.

EXAMPLE.—If 1 cord of wood cost $6.00, what will be the cost of $\frac{3}{4}$ of a cord?

$Solution.$—$6.00 $\times \frac{3}{4}$ = $4.50
3
$\overline{1800 \div 4} = 4.50$

What will $\frac{1}{3}$ of a cord cost?

$\frac{6}{1} \times \frac{1}{3} = \frac{6}{1} \div \frac{3}{1} = \$2.00$

MULTIPLICATION AND DIVISION. 57

EXAMPLES.

1. What will 360 barrels of flour come to at 5¼ dollars a barrel? At 1 dollar a barrel it would be 360 dollars; at 5¼ dollars, it would be 5¼ *times as much.*

```
                     360
                      5¼
                    ─────
5 times,            1800
¼ of a time,          90
                    ─────
```

Ans. $1890

Before we attempt to divide by a mixed number, such as 2½, 3¼, 5⅔, etc., we must explain, or rather *observe* the principle of division, namely: *That the quotient will be the same if we multiply the dividend and divisor by the same number.* Thus 24 divided by 8, gives three for a quotient. Now, if we double 24 and 8, or multiply them by any number whatever, and then divide, we shall still have 3 for a quotient. 16)48(3; 32)96(3, etc.

Now, suppose we have 22 to be divided by 5½; we may double both these numbers, and thus be clear of the fraction, and have the same quotient. 5½)22(4 is the same as 11)44(4.

How many times is 1¼ contained in 12? *Ans.* Just as many times as 5 is contained in 48. Th, 5 is 4 times 1¼, and 48 is 4 times 12. From these observations, we draw the following rule for dividing *by a mixed number.*

Rule.—*Multiply the whole number by the lower term of the fraction; add the upper term to the product for a divisor; then multiply the dividend by the lower term of the fraction, and then divide.*

How many times is $1\frac{1}{5}$ contained in 36? *Ans.* 30 times.

N. B. If we multiply both these numbers by 5, they will have the same *relation* as before, and a quotient is nothing but a relation between two numbers. After multiplication, the numbers may be considered as having the *denomination of fifths.*

How many times is $\frac{1}{4}$ contained in 12? *Ans.* 48 times.

One-fourth multiplied by 4, gives 1; 12, multiplied by 4, gives 48. Now, 1 in 48 is contained 48 times.

Divide 132 by $2\frac{3}{4}$. *Ans.* 48.
Divide 121 by $15\frac{1}{8}$. *Ans.* 8.
How many times is $\frac{3}{4}$ contained in 3? *Ans.* 4 times.

By a little attention to the relation of numbers, we may often contract operations in multiplication. A dead uniformity of operation *in all cases* indicates a mechanical and not a scientific knowledge of numbers. As a uniform principle, it is much easier to multiply by the **small numbers, 2, 3, 4, 5,** than by 7, 8, 9.

PERCENTAGE
AS APPLIED TO BUSINESS

The greater portion of all arithmetical calculations, as applied to every-day business transactions, being based upon percentage, it is important that the foregoing principles and illustrations be thoroughly mastered.

Percentage is the process of computing by the hundred.

Per cent., or *rate per cent.*, means by the hundred, and is represented by the character % instead of being written thus, 5%, 20%, 100%. Any % less than 1%, can be written in the form of a fraction, thus, $\frac{1}{4}$%, $\frac{3}{8}$%, or expressed decimally, thus, .002%, .0025%.

THE FIVE FACTS

To be considered in percentage are:

1st. *The Base.* 2d. *The Rate.* 3d. *The Percentage.* 4th. *The Amount.* 5th. *The Difference.*

The Base—Is the number upon which the *Percentage* is calculated.

The Rate—Is the number denoting the per cent. (or hundredths) of the *Base* taken, and is always used as the multiplier.

APPLICATION OF PERCENTAGE TO BUSINESS

PERCENTAGE.

The Percentage—Is the sum (in hundredths) obtained from multiplying the *Base by the Rate*.

The Amount—Is the *Base* increased by adding the *Percentage*.

The Difference—Is the *Base* diminished by subtracting the *Percentage*.

APPLICATION OF PERCENTAGE.

GIVEN—*The Base and Rate.*
To FIND—*The Percentage.*
RULE I.—*Multiply the* BASE *by the* RATE, *expressed decimally, and point off two places from the right.*

PROBLEM.

What is 9% (or the percentage) of $800?—

Process. Base, $800. Multiply by the rate.
         Rate, .09

B×R= Percentage, $72.00—Point off two figures from the right, and the result is the per cent. $72.

GIVEN—*The Base and Rate.*
To FIND—*The Amount.*
RULE II.—*To the* BASE *add the* PERCENTAGE.

PROBLEM.

What is the amount of $800, increased 9%?

Process. Base, $800. Plus the Percentage.
         Percentage,   72. Obtained under Rule I.

B+P=Amount, $872.

GIVEN—*The Base and Rate.*
To FIND—*The Difference.*
RULE III.—*From the* BASE *subtract the* PERCENTAGE.

PROBLEM.

What is the Difference (or proceeds) of $800, less 9%?

Process. Base, $800. Minus the Percentage.
Percentage, 72. Obtained under Rule I.

B—P=Difference, $728.—or Proceeds.

GIVEN—*The Base and Percentage.*
To FIND—*The Rate.*
RULE IV.—*Divide the* PERCENTAGE *by the* BASE, *expressed decimally.*

NOTE.—When cents are not shown in the percentage add two ciphers.

PROBLEM.

Bought of W. H. Sadler, invoice of Orton's Lightning Calculators, for $850, and sold them at a profit of $297.50—what % did I make? *Ans.* 35%.

Process.—Percentage divided by the Base.
Base, $850. Percentage, $297.50. P÷B=Rate.

Base. Percentage. Rate.
850)297.50(35%
   2550

    4250
    4250

PERCENTAGE.

Given—*Rate and Percentage.*

To Find—*The Base.*

Rule V.—*Divide the* Percentage *by the* Rate.

Note.—When cents are not shown in the percentage annex two ciphers.

PROBLEM.

Sold William Callen, Jr., invoice of Orton's Lightning Calculators, upon which I gained $297.50. Ascertaining my profit to be 35%—what was the amount or cost? *Ans.* $850.

Process.—Percentage divided by the rate. Percentage, $297.50. Rate, 35%. P÷R=Base.

```
       Rate. Percentage. Base.
       35)297.50(850.
          280
          ———
           175
           175
           ———
             0
```

Given—*Amount and Rate.*

To Find—*The Base.*

Rule VI.—*Divide the* Amount *by* 100, *added to the* Rate.

Note.—When cents are not shown in the amount annex two ciphers.

PROBLEM.

Sold John G. Scouten, invoice of Orton's Lightning Calculators, amounting to $1147.50, and made a profit of 35%. What did the books cost? *Ans.* $850.

Process.—Amount divided by 100 plus the rate.

Amount, $1147.50. Rate, 35%. 100 increased by the rate 35=1.35.

```
              Rate.   Amount.   Base or Cost.
            1.35)1147.50(850
                 1080
                 ————
                  675
                  675
                  ———
A÷100+R=Base.       0
```

GIVEN—*The Difference and Rate.*

TO FIND—*The Base.*

RULE VII.—*Divide the* AMOUNT *by* 100 *less the* RATE.

NOTE.—When cents are not shown in the difference annex two ciphers.

PROBLEM.

Invoice of Orton's Lightning Calculators sold William David, were damaged by water, and he was compelled to sell them for $552.50, thereby losing 35%. What did they cost? *Ans.* $850.

Process.—Difference divided by 100 less the rate.

Difference, $552.50. Rate, 35%. 100 less the rate 35=65.

```
             Rate. Difference. Base or Cost.
            65)552.50(850.
               520
               ———
                325
                325
                ———
D÷1.00—R=Base.     0
```

PROFIT AND LOSS

Are terms denoting the gain or loss arising from business transactions.

The preceding Rules under percentage are specially adapted to the majority of business transactions; we therefore call attention to their application:

Capital or *Cost* is treated as the *Base*.

Per cent. (%) of profit or loss is treated as the *Rate*.

Sum Gained or *Lost* is treated as the *Percentage*.

Selling Price is treated as the *Amount*.

Cost, less the Loss, is treated as the *Difference*.

SHORT METHODS in MERCHANDISING.

When the Rate is an *Aliquot part of* $1.00 *or* 100, instead of following *Rule I.* of Percentage the labor will be greatly abridged by applying the short method, as explained on page 48.

NOTE.—Aliquot parts of a number are such whole or mixed numbers as will divide it without a remainder. Thus 2, 2½, 3⅓, and 5 are aliquot parts of 10, being contained in it 5, 4, 3, and 2 times.

TABLE OF ALIQUOT PARTS.

Aliquot parts of	1	10	100 or $1.00	1000	1 Ton of 2000 lb.	1 ft. or 1 doz.	1 A.
One half is....	½	5	50	500	1000	6	80 sq. rd.
One third is...	⅓	3⅓	33⅓	333⅓	666⅔	4	
One fourth is..	¼	2½	25	250	500	3	40 sq. rd.
One fifth is....	⅕	2	20	200	400		32 "
One sixth is. .	⅙	1⅔	16⅔	166⅔	333⅓	2	
One eighth is.	⅛	1¼	12½	125	250		20 sq. rd.
One tenth is ..	1/10	1	10	100	200		16 "
One twelfth is etc.	1/12		8⅓	83⅓		1	

EXAMPLE.—Multiply 843 × 83⅓.

Process.—Since 83⅓ is 1/12 of 1000, 83⅓ times any number is 1/12 of 1000 times that number. Therefore, to multiply 843 by 83⅓ we simply multiply 843 by 1000, and divide the product by 12, the quotient will be the required product thus:

$$843 \times 1000 = 843000 \div 12 = 70250.$$

INTEREST

Is the sum paid for the use of money.

Simple Interest is interest on the principal only.

Annual Interest is simple interest on the principal, and on each year's interest from the date of its accruing until paid.

Compound Interest is interest allowed on interest and principal combined.

☞ Calculations in compound interest may be abridged by use of tables on page 209.

NOTE.—Interest may be compounded and added to the principal annually, semi-annually, or quarterly, as per agreement between lender and borrower.

Accurate Interest is interest calculated on the basis of 365 days to the year. It is reckoned by the usual methods, and $\frac{1}{73}$ of the sum deducted, except in case of leap year when $\frac{1}{81}$ of the interest is subtracted.

Legal Interest is the rate fixed according to law.

Usury is when a higher rate of interest is paid than is sanctioned by law.

NOTE.—See pages 297-298.

The Principal is the sum in use and upon which interest is paid.

---

* For a better understanding of the practical calculations of interest, the author refers to other portions of this work.

The Rate of Interest is the price paid for the use of one dollar.

The Amount is the principal with the accrued interest added.

As in Percentage there are five facts to be considered, viz.:

Principal, Rate per annum, *Interest, Time* and *Amount.*

APPLICATION OF PERCENTAGE.

The Principal is treated as the *Base.*

The Rate, or price paid per annum, is treated as the *Rate.*

The Interest is treated as the *Percentage.*

The Principal and Interest is treated as the *Amount.*

The Time is an additional element in Interest.

GIVEN—*Principal, Rate, and Time* (in days).

To FIND—*The Interest* at any rate per cent.

RULE I.—Time in days. *Multiply the* PRINCIPAL *by the* RATE, *and the product by the time (expressed in days); then divide the result by 36 and the quotient will be the* INTEREST *in mills, or* $\frac{1}{1000}$ *of* $1.00.

GIVEN—*Principal, Rate, and Time* (in months).

To FIND—*The Interest* at any rate per cent.

RULE II.—Time in months. *Multiply the* PRINCIPAL *by the* RATE, *and the product by the* TIME (*number of months*); *divide the result by* 12, *and the quotient will be the* INTEREST *in cents, or* $\frac{1}{100}$ *of* $1.00.

GIVEN—*Principal, Rate, and Time* (in years).
To FIND—*The Interest at any rate per cent.*
RULE III.—Time in years. *Multiply the* PRINCIPAL *by the* RATE, *and the product by the* TIME (*number of years*), *the result will be the* INTEREST *in cents, or* $\frac{1}{100}$ *of* $1.00.

NOTE.—The above rules are not specially recommended for general business use, but are here presented to call attention to the principle (percentage) upon which all interest calculations are based.

For Short Methods and Practical every-day rules the author refers to the portion of this work devoted exclusively to interest calculations.

GIVEN—*The Principal, Rate, and Time.*
To FIND—*The Interest and Amount.*
RULE.—*Calculate the* INTEREST *for the* TIME *at the stated* RATE, *and add to the* PRINCIPAL. *The product will be the* AMOUNT.

EXAMPLE.—What will $1000 amount to, invested for 8 months at 7% interest?

$P \times R \times T \div 12 =$ Interest.
Principal, $1000 + 46.67 = $1046.67 Amount.

Process.—Principal, 1000 × Rate, 7%.

```
           07
         ─────
         7000 × Time in months.
            8
         ─────
      12)56000  Product ÷ 12 = Interest
       $46.666                [$46.67.
```

GIVEN—*The Principal, Interest, and Time.*
TO FIND—*The Rate.*
RULE.—*Divide the stated* INTEREST *by the interest on the* PRINCIPAL, *for the* TIME *calculated at* 1% *per annum, the quotient will be the* RATE.

If $5000, invested for 1 year and 6 months, gains $525—what is the rate? *Ans.* 7%.

Process.—Principal, $5000 × Rate, 1½%, for 1 year and 6 months = $75. Interest at 1%—Stated Interest divided by $75 = the Rate.

Stated Int.

Interest for time @ 1% = $75)525(7% Rate.
525

GIVEN—*The Rate, Time, and Interest.*
TO FIND—*The Principal.*
RULE.—*Divide the stated* INTEREST *by the interest on one dollar for the stated* TIME, *at the stated* RATE.

NOTE.—When cents are shown in the interest **annex two,** and for mills three ciphers.

INTEREST. 71

EXAMPLE.—What principal will gain $525 interest in 1 year and 6 months, at 7%.

Process.—Interest on $1.00 for 1 year and 6 months, is .105, or 10½ cents.

Stated Int.
Interest on $1.00 = .105)525000(5000 Principal.
　　　　　　　　　　525
　　　　　　　　　　———
　　　　　　　　　　　000

GIVEN—*The Rate, Time, and Amount.*
To FIND—*The Principal.*
RULE.—*Divide the* AMOUNT *by* 100, *plus the interest on one dollar for the stated* TIME, *at the stated* RATE.

NOTE.—When cents are shown in the interest annex two, and for mills three ciphers.

EXAMPLE.—What principal will amount to $5525, in 1 year and 6 months, at 7%?

Process.—Interest on $1.00 for 1 year and 6 months, at 7% = .105, or 10½ cents.

100 + .105 = 1.105)$525.000(5000 Principal.
　　　　　　　　　　525
　　　　　　　　　　———
　　　　　　　　　　　000

GIVEN—*Principal, Rate, and Interest.*
To FIND—*The Time.*
RULE.—*Divide the stated* INTEREST *by the interest on the* PRINCIPAL, *for one year, at the stated* RATE.

NOTE.—The integer or whole number in the quotient will be the time in years. *For months,* multiply the decimal or remainder by 12, and divide as before, the quotient will be the time in months. *For days,* multiply the decimal or remainder in months by 30, and divide again, the quotient will be the time in days.

EXAMPLE.—In what time will $5000 amount to $5455?

Process.—Principal, $5000. Interest 1 year at 6% = $300.

Stated Interest, $455 ÷ 300 = Time, 1 year, 6 months, 6 days.

```
300)455(1 year.
    300
    ---
    155
     12
300)1860(6 months.
    1800
    ----
      60
      30
300)1800(6 days.
    1800
```

TEST EXAMPLE.

In what time will $3000, at 7%, amount to $3570.50? *Ans.* 2 years, 8 months, 18 days.

NOTE.—When cents are shown in the stated interest annex two ciphers to the interest on the principal for one year, providing cents are not shown, and *vice versa.*

DISCOUNTS

Is the percentage off or allowance made for the payment of money before maturity.

COMMERCIAL DISCOUNT.

In this connection the term discount is used without reference to time. It is the sum or percentage deducted from the *List* or asking price of goods.

It is the allowance or deductions made from Invoices or Bills purchased, in consideration for prompt or *cash* payment.

NOTE.—Certain goods usually sold on credit may be bought for less price, providing cash settlements are made. The sum or abatement from the credit price or terms, such as, 2½, 5, or 6% off, is termed discount.

Again, on various classes of articles the retail price is fixed by the publisher or manufacturer, and certain deductions are allowed to importers or wholesale buyers, which is given in the form of a per cent. off, such as, 25, 33½, and 40%, with further allowances for *Net Cash* payment.

Net Price of an article is the selling or asking price, less the discount.

Net Proceeds, or cash value of a bill, is its face with the discount deducted.

APPLICATION OF THE RULE OF PERCENTAGE

Base—The selling price or face of bill.

Rate—The rate per cent. of deduction.

Percentage—The discount or amount of deduction.

RULE I.—*Multiply the selling price or face of the bill by the rate per cent. of deduction, and the product will be the Commercial discount.*

RULE II.—*From the selling price or face of the bill deduct the commercial discount, and the difference will be the Net Price, Cash Value, or Net Proceeds.*

NOTE.—The practical application of the above rules has, previously, been so fully illustrated that examples here are not deemed necessary.

TRUE DISCOUNT

Is the difference between the face of the debt and its present worth or value, therefore it is evidently the interest on the present worth from date to the time of maturity.

NOTE.—Every debt or note due at some future time, without interest, has some existing value *now*, and that value is termed Present Worth; therefore,

Present Worth is such a sum as being placed at interest to the date of maturity as will amount to the stated debt.

APPLICATION OF PERCENTAGE.

The Present Worth is treated as the *Base.*

The Debt or *Face of Bill* is treated as the *Amount.*

The True Discount is treated as the *Difference.*

PERCENTAGE.

TO ASCERTAIN THE PRESENT WORTH.

RULE.—*Divide the amount of the debt or face of bill by* 100 *plus the interest on* $1.00 *for the given time, at the stated rate.*

NOTE.—When cents are shown in the interest annex two, for mills three, ciphers, to the debt or bill.

TO ASCERTAIN THE TRUE DISCOUNT.

RULE.—*From the debt or face of bill subtract the present worth.*

EXAMPLE.—What is the present worth of $618—Note due in 6 months, at 6%?

Process.—Int. on $1.00 6 mos. @ 6% is .03
.03+100=1.03)618.00(600.—Present worth.
          618
         —————
            00

618—600=18.—True discount.

Proof.—1.00×600=$600.
       618—600= 18.
                  ———
Face of note, $618.

BANK DISCOUNT

Is the *Interest* paid in advance, or deducted from the face of a note or time draft.

NOTE.—Should the paper offered for discount bear interest, bank discount is the interest on the amount due at maturity instead of on the face. In discounting, the time is reckoned by days, and the basis of calculation 360 days to the year. In discounting paper, banks include the day on which the note is discounted and the day on which it matures.

Discount is the sum deducted from the face of a note or acceptance, which is the interest for the number of days from date of discount to maturity.

Proceeds is the sum given or amount of the note or acceptance, discounted, less the interest.

Maturity of a note is the time or date it becomes due, including days of grace.

Days of Grace are the three days allowed by law for payment, after the expiration of the time specified in the note.

Protest is the formal legal notice made by a Notary Public, notifying the maker and endorsers of the non-acceptance or payment of paper for which they are held liable.

NOTE.—A protest for non-payment must be made on the last of three days of grace, unless that day should occur on Sunday, or legally authorized holiday, in which case protest must be made on the day previous.

Non-Protest.—In case of non-protest wherein there are endorsers to commercial paper, they are legally released, and the holder can only look to the maker for payment.

Protest Waived.—Consent of drawers or endorsers to hold themselves responsible for payment without the necessity of protest.

COMMISSION.

The rate of commission or brokerage in generality of cases is established by custom, ranging from ⅛% to 7%. A commission merchant generally gets 2½% for selling, and an additional 2½% if he guarantees the payment.

Commission—Is the sum paid by the principal to an agent for selling goods or property, or collecting money.

Consignment—Goods sent to a commission merchant to be sold.

Consignor—The party sending the goods, or shipper.

Consignee—The party to whom the goods are sent.

Proceeds—The sum remaining after all expenses are paid.

Account Sales—Consignees' written statement to the consignor, showing at what price the goods were sold, the expenses, and the net proceeds.

Guarantee—Pledge or security given by the commission merchant for all goods sold on credit.

Broker—One who sells or purchases goods, stock, etc., by direction of another, without having them in his possession.

Brokerage—The sum paid a broker for his services.

The principles and workings of percentage involved in commission and brokerage are the same as those heretofore treated.

CORRESPONDING TERMS.

The *Base* is the amount of sales, investments, or collecting.

The *Rate* is the per cent. allowed for services.

The *Percentage* is the *Commission* or *Brokerage*.

The *Amount* or *Difference* is the *Net Proceeds*.

RULE I.—*To find the Commission or Brokerage, multiply the Base by the Rate.*

Rule II.—*To find the Rate, divide the Commission by the Base.*

Rule III.—*To find the Base, divide the Commission by the Rate.*

Note.—Wherein a certain sum is supplied a broker for investment or purchases, from which the pay for his commission is to be taken; commission on his own commission not allowed, we have the following

Rule.—*Divide the sum supplied by* 100 *plus the rate % of commission, the quotient will be the Net Proceeds; this sum subtracted from the Amount will give the Commission.*

Example.—James G. Moulton remits a broker $10,000 with instructions to invest in cotton, his commission, $2\frac{1}{2}\%$, which is to be deducted—what is the amount of cotton purchased? What is his commission?

$$\begin{array}{rl} \text{Working amount to be invested,} = & 100 \ \% \\ \text{Commission on sum,} = & 2\frac{1}{2} \\ \hline \text{Total on purchase,} = & 102\frac{1}{2}\% \end{array}$$

Sum furnished, $10,000÷1.025=$9756.10 Inv.
$10,000−9756.10=243.90 Commission.
$9756.10×.025=243.90 Commission.

TEST EXAMPLE.

How many bushels of corn can be purchased for $3485—the market price being 68 cents per bushel, and your agent's commission for purchasing $2\frac{1}{2}\%$? *Ans.* 5000 bushels.

INSURANCE

Is a contract issued by companies, wherein they agree for a certain consideration to indemnify the owner or holder of certain property against loss or damage by fire or shipwreck, etc.

The Underwriter is the Insurance Agent who acts for the company.

The Insured is the party asking for protection, and in whose favor the policy is issued.

The Policy is the written contract issued by the company, describing the property, amount of risk, and conditions of indemnity.

The Rate or per cent. of Premium is the cost of $100 of insurance.

The Premium is the amount paid the company for insurance, and is generally calculated at a certain per cent. on the amount of insurance.

PERCENTAGE AS APPLIED TO INSURANCE.

The Amount is treated as the *Base*.
The per cent. of Premium is treated as the *Rate*.
The Premium is treated as the *Percentage*.

GIVEN—*Amount of Insurance and Rate.*
TO FIND—*The Premium.*
RULE.—*Amount × by the Rate = Premium.*

GIVEN—*Premium and Rate of Insurance.*
TO FIND—*The Amount.*
RULE.—*Premium expressed in cents ÷ by the Rate = Amount.*

GIVEN—*Amount and Premium.*
TO FIND—*The Rate.*
RULE.—*Premium expressed in cents ÷ by the Amount = Rate.*

INVESTMENTS.

CAPITAL AND STOCKS.

Capital is money invested in business or private enterprises, conducted under individual or co-partnership management.

Capital Stock is money or property invested by sundry persons in manufacturing, railroading, banking, etc., and is generally divided into certificates or shares of $100 each.

The management of such enterprises is controlled by a Board of Directors, from among whose number executive officers are elected or appointed.

Certificates of Stock are official documents issued by the corporation or company, representing a certain number of shares of the joint capital to which the holder is entitled.

The Par Value of stocks is the sum or nominal value for which they were issued, as expressed on their face.

The Market Value is the sum for which they can be sold.

Stocks are at Par when their market value is the same as their face.

Stocks are below Par when their market value is less than their face.

Stocks are above Par when their market value is in excess of their face value.

NOTE.—Shares representing $100 each, when quoted at $100, are worth par; when at $110 or over $100 are above par, and when at $90 or less than $100 are below par.

Market quotations of stocks are generally quoted at a certain per cent. above or below the par value.

The value of stocks depend upon the success and prosperity of the business they represent, and per cent of dividend declared.

Assessment is the sum called for from the stockholders to make up any deficiency or losses that may arise in conducting the business.

Dividend is the sum paid the stockholders, and is a division of the profits of the company.

Assessments and Dividends are calculated upon a certain per cent. of the par value of the stock.

INVESTMENTS. 83

Brokerage.—The party buying and selling stocks is called a Broker or Stock Jobber, and the compensation received for his services is termed Brokerage. The usual rate of brokerage is ⅛ to ¼ per cent. of the par value of stock purchased or sold.

PERCENTAGE.

The majority of business transactions that arise in connection with stocks may be readily calculated by applying the principles of percentage heretofore shown, as an examination of the following illustrations will show:

To ascertain the Cost, *including* Brokerage.

Rule.—*To the market value of one share add the brokerage, and multiply by the number of shares.*

Example.—What will 100 shares of Baltimore & Ohio Railroad stock cost, market value 127⅜, brokerage ⅛%?

PROCESS.

127⅜ (Cost of 1 share) + ⅛ (Brokerage) = 127½ Cost of 1 share.
127½ × 100 (Number of shares,) = $12,750 Cost.

To ascertain the number of Shares.

Rule.—*To the market value of one share add the brokerage (if any), and divide the sum to be invested by the amount thus obtained.*

EXAMPLE.—How many shares of Baltimore & Ohio Railroad stock can be purchased for $12,750? Market value 127¾, brokerage ⅛%. *Ans.* 100 shares.

PROCESS.

127¾ (Market value) + ⅛ (Brokerage) = 127½ Cost of 1 share.
$12,750 ÷ 127½ = 100 Shares.

To ascertain amount of INVESTMENT.

RULE.—*Divide the stated income by the income on one share* (which will give the number of shares required). *Multiply the number of shares by the cost per share, and the product will be the required investment.*

EXAMPLE.—How much capital must be invested in New York Central Railroad stocks @ 110, which pay semi-annual dividends of 6%, to realize an income of $900 per annum? *Ans.* $8250.

PROCESS.

$900 (Desired Income) ÷ $12 (Income on 1 share) = 75 No. of shares.
$110 (Cost of 1 share) × 75 (Number of shares) = $8250 Investment.

To ascertain the RATE % *of income realized from investments.*

RULE.—*Divide the annual dividend or income on one share by the cost per share.*

EXAMPLE.—If Railroad shares paying annual dividends of 10% command a premium of 25%—what per cent. of income will be realized from investing in said shares? *Ans.* 8%.

INVESTMENTS.

PROCESS.

$10 (Income from 1 share) ÷ $125 (Cost of 1 share) = 8%.

To ascertain at what price stocks must be bought to produce a certain INCOME.

RULE.—*Divide the dividend or income on one share by the desired rate of interest or income.*

EXAMPLE I.—What amount of premium must stocks bring, paying annual dividends of 12%, to net 9% income to the investor? *Ans.* 33⅓% Premium.

PROCESS.

$12 (Income from 1 share)÷9% (Req. int.)=$133⅓ Value of 1 share. 133⅓—100 Par value=33⅓% Premium.

EXAMPLE II.—At what price must stock producing annual dividends of 6% be bought so as to net the investor 9%? *Ans.* 33⅓% discount.

PROCESS.

$6 (Income from 1 share)÷9% (Req. int.)=$66⅔ Value of 1 share. $100 (Par value of 1 share)—66⅔ (Market value)=33⅓% Discount.

TEST EXAMPLE.

At what price must stock of the par value of $50 per share, which pays annual dividends of $3 per share, be bought to produce an income of 7½%? *Ans.* $40.

TABLE FOR INVESTORS.

The following Table shows the rate per cent. of Annual Income from Bonds bearing 5, 6, or 7 per cent. interest, and costing from 50 to 125.

Purchase Price.	5%	6%	7%	Purchase Price.	5%	6%	7%
50	10.00	12.00	14.00	88	5.68	6.81	7.94
51	9.80	11.76	13.72	89	5.61	6.74	7.86
52	9.61	11.53	13.46	90	5.55	6.66	7.77
53	9.43	11.32	13.20	91	5.49	6.59	7.69
54	9.25	11.11	12.96	92	5.43	6.52	7.60
55	9.00	10.90	12.72	93	5.37	6.45	7.52
56	8.92	10.70	12.50	94	5.31	6.38	7.44
57	8.77	10.52	12.27	95	5.26	6.31	7.36
58	8.62	10.34	12.06	96	5.20	6.25	7.29
59	8.47	10.16	11.86	97	5.15	6.18	7.21
60	8.33	10.00	11.66	98	5.10	6.12	7.14
61	8.19	9.83	11.47	99	5.05	6.06	7.07
62	8.06	9.67	11.29	100	5.00	6.00	7.00
63	7.93	9.52	11.11	101	4.95	5.94	6.93
64	7.81	9.37	10.93	102	4.90	5.88	6.86
65	7.69	9.23	10.76	103	4.85	5.82	6.79
66	7.57	9.09	10.60	104	4.80	5.76	6.72
67	7.46	8.95	10.44	105	4.76	5.71	6.66
68	7.35	8.82	10.29	106	4.71	5.66	6.60
69	7.24	8.69	10.14	107	4.67	5.60	6.54
70	7.14	8.57	10.00	108	4.62	5.55	6.48
71	7.04	8.45	9.85	109	4.58	5.50	6.42
72	6.94	8.33	9.72	110	4.54	5.45	6.36
73	6.84	8.21	9.58	111	4.50	5.40	6.30
74	6.75	8.10	9.45	112	4.46	5.35	6.25
75	6.66	8.00	9.33	113	4.42	5.30	6.19
76	6.57	7.89	9.21	114	4.38	5.26	6.14
77	6.49	7.79	9.00	115	4.35	5.21	6.08
78	6.41	7.69	8.97	116	4.31	5.17	6.03
79	6.32	7.59	8.86	117	4.27	5.12	5.98
80	6.25	7.50	8.75	118	4.23	5.08	5.93
81	6.17	7.40	8.64	119	4.20	5.04	5.88
82	6.09	7.31	8.53	120	4.16	5.00	5.83
83	6.02	7.22	8.43	121	4.13	4.95	5.78
84	5.95	7.14	8.33	122	4.09	4.91	5.73
85	5.88	7.05	8.23	123	4.06	4.87	5.69
86	5.81	6.97	8.13	124	4.03	4.83	5.65
87	5.74	6.89	8.04	125	4.00	4.80	5.60

INTEREST

Definition of Terms

Interest is premium paid for the use of money, goods, or property.

It is computed by percentage—a certain per cent. on the money being paid for its use for a stated time. The money on which interest is paid is called the PRINCIPAL.

The per cent. paid is called the RATE; the principal and interest added together is called the AMOUNT.

When a rate per cent. is stated, without the mention of any term of time, the time is understood to be 1 year.

The first important step in the calculation of simple interest is the arranging of the time for which it is computed. The student must study the following Propositions carefully, if he would be expert in this important and useful branch of business calculations:

PROPOSITION 1.

If the time consists of years, multiply the principal by the rate per cent., and that product by the number of years.

EXAMPLE 1.—Find the interest of $75 for 4 years at 6 per cent.

Operation.

```
   $75            The decimal for 6 per cent. is
   .06            06.  There being two places of
   ----           decimals in the multiplier, we
   4.50           point off two in the product.
   4
   -----
  $18.00 Ans.
```

PROPOSITION 2.

If the time consists of years and months, reduce the time to months, and multiply the principal by the rate per cent. and number of months together, and divide the result by 12.

NOTE.—The work can always be abbreviated at 4, 6, 8, 9, 12, and 15 per cent., by canceling the per cent., or time, or principal, with the common divisor 12.

INTEREST.   89

EXAMPLE 2.—Find the interest of $240 for 2 years and 7 months at 8 per cent.

First method.
  Principal, $240
  Per cent.,   .08
                  ———
  In. for 1yr., 19.20
  2yrs.+7mos., 31mos.
                ———
        12)595.20
            ———
       $49.60 *Ans.*

Second method: by cancellation.
       | ~~240~~—20
  ~~12~~ | 8   rate.
       | 31   time
          ———
      49.60 *Ans.*

The operation by canceling is much more brief. We simply place the principal, rate, and time, on the right of the line, and 12 on the left; then we cancel 12 in 240, and the quotient 20 multiplied with 8 and 31 gives the interest at *once*.

NOTE.—After 12 is canceled the product of the remaining numbers is *always* the interest.

PROPOSITION 3.

If the time consists of years, months, and days, reduce the years to months, add in the given months, and place one-third of the days to the right of this number, which we multiply by the principal and rate per cent., and divide by 12, as before, or cancel and divide by 12 before multiplying.

EXAMPLE 3.—Find the interest of $231 for 1 year, 1 month, and 6 days, at 5 per cent.

First method.
Principal, $231
Per cent.,   .05
                    ─────
In. for 1yr.,   11.55
1yr.+1mo.+6da.,   13.2mo.
                    ─────
            12)152.460
                    ─────
            $12.705 Ans.

Second method:
by cancellation.
           | 231   prin.
        12 | 5   rate.
           | 1̶3̶2̶ —11
           ─────
           $12.705 Ans.

By the second method we cancel 12 in 132, and multiply the quotient 11 by 5 and 231.

NOTE.—When the principal is $, and the time is in years or months, the interest is in cents; if the time is in years, months, and days, the interest is in mills, unless the days are less than 3, in which case it would be in cents, as before.

NOTE.—The reason we divide the days by 3 is because we calculate 30 days for a month, and dividing by 3 reduces the days to the tenth of months

NOTE.—The three preceding propositions will work any note in interest for any time and at any given rate per cent.

How to Avoid Fractions in Interest.
PROPOSITION 4.

If, when the time consists of years, months, and days, are not divisible by 3, you can divide the days by 3, and annex the mixed number as in Proposition

INTEREST. 91

3, or if you wish to avoid fractions, you can reduce the time to interest days, and multiply the principal, rate and days together, and divide the result by 36 or its factors, 4×9.

NOTE.—In this case as in the preceding, the work can almost always be contracted by dividing the rate or time or principal with the divisor 36.

NOTE.—We use the divisor 36, because we calculate 360 interest days to the year. We discard the 0, because it avails nothing to multiply or divide by.

EXAMPLE 4.—Find the interest of $210 for 1 year, 4 months, and 8 days, at 9 per cent.

Year. Months. Days.
 1   4   8 = 16.2⅔ months or 488 days.

Operation By Prop. 3.	Operation By Prop. 4.
$210	$210
.9	9
18.90	18.90
16.2⅔	488
12)307440	36)922320
$25.620 *Ans.*	$25.620 *Ans.*

We will now work the example by cancellation to show its brevity.

Operation by Cancellation.

Time 488 days.

$$\begin{array}{c|c} & 210 \\ 4\text{--}36 & 9 \\ & 488 \quad 122 \\ & 122 \\ & 210 \\ \hline & \$25.620 \end{array}$$

Now cancel 9 in 36 goes 4 times, then 4 into 488 goes 122. Now multiply remaining numbers together, thus, 210×122 and we have the interest at once.

When the days are not divisible by 3 we reduce the whole time to days; then we place the principal rate and time on the right of the line. Now, because the time is in days, we place 36, on the left of the line for a divisor. (*If the time was months we would place 12 on the left.*)

NOTE.—A very short method of reducing time to interest days is to multiply the years by 36; add in 3 times the number of months and the tens' figure of the days, and annex the unit figure; but if the days are less than 10 simply annex them.

EXAMPLE 1.—Reduce 1 year, 2 months, and 6 days, to days.

Years. Months. Days.
36×1+3×2=42 annex 6=426 *Ans.*

EXAMPLE 2.—Reduce 2 years, 3 months and 17 days to interest days.

Years.　M'ths.　Days.　　　Days.
$36 \times 2 + 3 \times 3 + 1 = 82.$ annex $7 = 827$ days. *Ans.*

NOTE.—The student should commit to memory the multiplication of the number 36 up as far as 9 times 36, and then he can reduce almost instantly years, months, and days, to days.

SIMPLE INTEREST BY CANCELLATION.

RULE.—*Place the principal, time, and rate per cent. on the right hand side of the line. If the time consists of years and months, reduce them to months, and place 12 (the number of months in a year) on the left hand side of the line. Should the time consist of months and days, reduce them to days or decimal parts of a month. If reduced to days, place 36 on the left. If to decimals parts of a month, place 12 only as before.*

Point off two decimal places when the time is in months, and three decimal places when the time is in days.

NOTE.—If the principal contains cents, point off four decimal places when the time is in months, and five decimal places when the time is in days.

NOTE.— *We place 36 on the left because there are 360 interest days in a year. (Custom has made this lawful.)*

EXAMPLE 1.—What is the interest on $60 for 117 days at 6 per cent?

Operation.

Here 117 × 0 must be the 36 answer.

$$\begin{array}{r} \cancel{6}0 \\ \cancel{6} \\ 117 \\ \hline \$1.170 \ \textit{Ans.} \end{array}$$

Both sixes on the right cancels 36 on the left, and we have nothing left to divide by.

In this case we point off three decimal places because the time is in days. If the time had been 117 months, we would have pointed off but two decimal places.

EXAMPLE 2.—What is the interest of $96.50 for 90 days at 6 per cent?

Operation.

6—36 | 96.50
     | 9̶0̶—-15
     | 6̶

9650
15
―――
1.44.750 *Ans.*

Now cancel 6 in 36 and the quotient 6 into 90, and we have no divisor left. Hence 15 × 96.50 must be the answer.

NOTE—As there are cents in the principal, we point off five decimals; three for days and two for cents Pay no attention to the decimal point until the close of the operation.

SIMPLE INTEREST BY CANCELLATION. 95

EXAMPLE 3.—What is the interest of $480 for 361 days at 6 per cent?

```
       | 4̶8̶0̶—80        361
6—3̶6̶   | 361            80
       | 6̶             ─────
                       $28.880 Ans.
```

Now cancel 6 in 36 and the quotient 6 into 480, and we have no divisor left. Hence 80×361 must be the answer.

EXAMPLE 4.—What is the interest of $720 for 9 months at 7 per cent?

```
       | 7̶2̶0̶—60        60
  1̶2̶   | 9              9
       | 7             ─────
                        540
                          7
                       ─────
                       $37.80 Ans.
```

Now cancel 12 in 720 there is nothing left to divide by. Hence 60×9×7 must be the answer.

N. B. When interest is required on any sum for days only, it is a universal custom to consider 30 days a month, and 12 months a year; and, as the unit of time is a year, the interest of any sum for *one day* is $\frac{1}{360}$, what it would be for a year. For 2 days, $\frac{2}{360}$, etc.; hence if we multiply by the days, we must divide by 360, or divide by 36 and save labor. The old form of this method was to place 360, or 12 and 30, on the left of the line but using 36 is much shorter.

WHEN THE DAYS ARE NOT DIVISIBLE BY THREE.

NOTE.—When the time consists of months and days, and the days are not divisible by three, *reduce the time to days.*

EXAMPLE 5.—What is the interest of $960 for 11 months and 20 days at 6 per cent?

$$\text{Operation.} \quad \overset{\text{Months. Days.}}{11 \quad 20} = 350 \text{ days.}$$

```
                   Months.  Days.
  Operation.        11     20 = 350 days.
                  | 960—160        350
      6 —36       | 350             160
                  | 6             _____
                                 $56.000
```

Now cancel 6 in 36 and the quotient 6 into 960, and we have no divisor left. Hence 160×350 must be the answer.

EXAMPLE 6.—What is the interest of $173 for 8 months and 16 days at 9 per cent?

```
                   Months.  Days.
  Operation.        8     16 = 256 days.
                  | 173            173
      9 —36       | 9               64
                  | 256—64        _____
                                 $11.072 Ans.
```

Now cancel 9 in 36 and the quotient 4 into 256, and we have no divisor left. Hence 64×173 must be the answer.

N. B. Let the pupil remember that this is a general and universal method, equally applicable to any per cent. or any required time, and all other rules must be reconcilable to it; and, in fact, all other rules are but modifications of this.

SIMPLE INTEREST BY CANCELLATION

EXAMPLE 7.—What is the interest on $1080 for 7 months and 11 days at 7 per cent?

$$\begin{array}{cc} \text{Months.} & \text{Days.} \\ 7 & 11 = 221 \text{ days.} \end{array}$$

Operation.

$$\require{cancel}\cancel{36}\,\bigg|\begin{array}{c}\cancel{1080}\text{—}30 \\ 221 \\ 7\end{array}$$

$$\begin{array}{r} 221 \\ 30 \\ \hline 6630 \\ 7 \\ \hline \$46,410 \ Ans. \end{array}$$

Now cancel 36 in 1080 and we have no divisor left, hence $30 \times 221 \times 7$ must be the answer.

WITH MORE DIFFICULT TIME AND RATE PER CENT.

EXAMPLE 8.—What is the interest of $160 for 19 months and 23 days at $4\frac{1}{2}$ per cent?

$$\begin{array}{cc} \text{Months.} & \text{Days.} \\ 19 & 23 = 593 \text{ days.} \end{array}$$

Operation.

$$8\text{—}\cancel{36}\,\bigg|\begin{array}{c}160\text{—}20 \\ 593 \\ \cancel{4\frac{1}{2}}\end{array}$$

$$\begin{array}{r} 593 \\ 20 \\ \hline \$11.860 \ Ans. \end{array}$$

Now cancel $4\frac{1}{2}$ in 36 and the quotient 8 into 160 for have no divisor left, hence 20×593 must be the interest.

WHEN THE DAYS ARE DIVISIBLE BY THREE.

RULE.—*Place one-third of the days to the right of the months, and place 12 on the left of the line.*

EXAMPLE 11.—What is the interest of $350 for 3 years 7 months and 6 days at 10 per cent?

    Years. Months. Days.
     3   7   6 = 43.2 months.

Operation.
```
        | 350                      350
   1̶2̶  | 4̶3̶2̶—36                    36
        |  10                    ——————
                                 12600
                                    10
                                ——————
                                $126.000 Ans.
```

Now cancel 12 in 432 and we have no divisor left. Hence $350 \times 36 \times 10$ equals the interest.

EXAMPLE 12.—What is the interest of $241 for 13 months and 9 days at 8 per cent?

    Months. Days.
     13   9 = 13.3 months.

Operation.
```
          | 241                    241
   3–1̶2̶   | 13.3                   133
          | 8̶—2                   ——————
                                  32053
                                      2
                                ——————
                                3)64106
                                ——————
                                $21.368⅔ Ans.
```

In this example I canceled 8 and 12 by 4, and then multiplied all on the right of the line and divided by 3. If I could have divided by 3 before multiplying I would have saved labor, but when the numbers are prime the whole work must be *literally* done.

CLOSING REMARKS.—We have now fully explained the canceling system of computing interest. Any and every problem can be stated by this method, and the beauty and simplicity of the system ranks it high among the most important abbreviations ever discovered by man. As we have before remarked, at 6, 4, 8, 9, 12, 15, and 4½ per cents., every problem in interest can be canceled, besides a great many can be abbreviated at 5, 7, and other per cents.; and after the problem has been stated and we find that we can not cancel, what have we done? We have simply stated the problem in its *simplest* and *easiest* form for working it by *any* other method. Hence we have a decided advantage of *all* notes that will cancel, and if we can not cancel we have stated the problem in its correct and proper form for going through the whole work; but it is only when the principal, time, and rate per cent. are *all* prime, that the WHOLE work must be LITERALLY done. At 6 per cent. we can cancel through, and 6 is the rate *most commonly used.*

SHORT PRACTICAL RULES,

DEDUCED FROM THE CANCELING SYSTEM,

For calculating interest at 6 per cent., either for months, or months and days.

To find the interest for months at 6 per cent.

RULE.—*Multiply the principal by half the number of months, expressed decimally as a per cent.; that is, for 12 months, multiply by .06; for 8 months, multiply by .04.*

NOTE 1.—It is obvious that if the rate per cent. were 12, it would be 1 per cent. a month; if, therefore, it be 6 per cent., it will be a half per cent. a month; that is, half the months will be the per cent.

NOTE 2.—If any other per cent. is wanted you can proceed as above, and then multiply by the given rate per cent. and divide by 6, and the quotient is the interest.

1. What is the interest of $368 for 8 months?

$368
.04=half the months.
———
$14.72=*Ans.*

NOTE 3.—When the months are not even; that is, will not divide by 2, *multiply one-half the principal* by the whole number of months, expressed deci mally.

To find the interest of any sum at 6 per cent. per annum for any number of months and days.

RULE.—*Divide the days by 3 and place the quotient to the right of the months; one-half of the number thus formed multiplied by the principal, or one-half of the principal multiplied by this number, will give the interest—pointing off three decimal places when the principal is $.*

2. What is the interest of $76 for 1 year, 6 months, and 12 days, at 6 per cent?

```
   Years. Months. Days.
     1      6     12 = 18.4 months—half 9.2.
    $76            Or,        184
    9.2                        38 = half prin.
   ──────                    ──────
  $6.992 Ans.               $6.992 Ans.
```

NOTE.—Dividing the days by 3 reduces them to the tenth of months.

To find the the interest of any sum at 6 per cent. per annum for any number of days.

RULE.—*Divide the principal by 6 and multiply the quotient by the number of days; or divide the days by 6 and multiply the quotient by the principal, pointing off three decimal places when the principal is $.*

NOTE.—Always divide 6 into the number that

will divide without a remainder; if neither one will divide, multiply the principal and days together and divide the result by 6.

3. What is the interest of $240 for 18 days at 6 per cent?

$18 \div 6 = 3$ $240 \div 6 = 40$
    240 *Or,* $40 = \frac{1}{6}$ of prin.
   $3 = \frac{1}{6}$ of the days. 18

   0.720 *Ans.* 0.720 *Ans.*

4. What is the interest of $1800 for 72 days at 6 per cent.

   1800 *Or,* $300 = \frac{1}{6}$ of prin.
   $12 = \frac{1}{6}$ of the days. 72

   21.600 *Ans.* 21.600 *Ans.*

Useful Suggestions to the Accountant in Computing Interest at 6 per cent.

If the principal is divisible by 6, *always reduce the time to days;* then multiply the number of days by one-sixth of the principal.

EXAMPLE.

5. Find the interest of $240 for 1 year, 5 months and 17 days, at 6 per cent.

       $6)240$ 1yr, 5mos., 17da. $= 527$ days
          ——   Multiplied by   40
$\frac{1}{6}$ of prin. $= 40$ ———
                                   21.080 *Ans.*

If the principal is divisible by 3, multiply one-third of the principal by one-half of the days.

6. What is the interest of $210 for 80 days at 6 per cent.?

$70 = ⅓ of the principal.
40 = ½ of the days.

$2.800 *Ans.*

When the Rate of Interest is 4 per cent

RULE.—*Multiply the principal by one-third the number of months, or by one-ninth the number of days, and the product is the interest.*

NOTE.—This principle is also deduced from the canceling method of computing interest; the student can readily see that 4 is ⅓ of 12 and ⅑ of 36.

When the Rate of Interest is 9 per cent.

RULE.—*Multiply the principal by three-fourths the number of months, or one-fourth the number of days, or vice versa.*

BANKS AND BANKING.

A BANK is an institution established under legal charter for the purpose of trafficking in money.

They issue notes payable to bearer on demand, which circulate as money, receive money on deposit, and make loans.

MONEY DEPOSIT in Banks is generally subject to the order of the depositor by check.

A BANK CHECK is the written order of the depositor for the payment of money.

NOTES.—In discounting business-paper it is customary for banks to calculate or take off the interest for the actual number of days from the day of *discount* to date of *maturity*, both days inclusive.—See Bank Discount, page 76.

"How to Transact Business with Banks."—See page 281.

The endorser of a note incurs all the obligations of such endorsement, even though he may be ignorant of the law at the time.

Many a man has been reduced from affluence to poverty by merely writing his name on the back of a note, "*just to accommodate a friend.*"

BANKERS' METHOD OF COMPUTING INTEREST.

AT 6 PER CENT. FOR ANY NUMBER OF DAYS.

RULE.—*Draw a perpendicular line, cutting off the two right hand figures of the $, and you have the interest of the sum for 60 days at 6 per cent.*

NOTE.—The figures on the left of the line are $, and those on the right are decimals of $.

EXAMPLE 1.—What is the interest of $423 60 days at 6 per cent.?

$423 = the principal.
$4 | 23 cts. = interest for 60 days.

NOTE.—When the time is more or less than 60 days, first get the interest for 60 days, and from that to the time required.

EXAMPLE 2.—What is the interest of $124 for 15 days at 6 per cent.?

          Days.   Days.
          15 = ¼ of 60

$124 = principal.
4)1 | 24 cts. = interest for 60 days.
     | 31 cts = interest for 15 days.

EXAMPLE 3.—What is the interest of $123.40 for 90 days at 6 per cent.?

$$\overset{\text{Days. Days. Days.}}{90=60+30}$$

$123.40 = principal.
2)1 | 2340 = interest for 60 days.
    | 6170 = interest for 30 days.
———

Ans. $1 | 851 = interest for 90 days.

EXAMPLE 4.—What is the interest of $324 for 75 days at 6 per cent.?

$324 = principal. $\overset{\text{Days. Days. Days.}}{75=60+15}$
4)3 | 24 cts. interest for 60 days.
    | 81 cts. interest for 15 days.
———

Ans. $4 | 05 cts. interest for 75 days.

REMARKS.—This system of Computing Interest is very easy and simple, especially when the days are aliquot parts of 60, and one simple division will suffice. It is used extensively by a large majority of our most prominent bankers; and, indeed, is taught by most all Commercial Colleges as the shortest system of computing interest.

Method of Calculating at Different Per Cents.

This principle is not confined alone to 6 per cent. as many suppose who teach and use it. It is their custom *first* to find the interest at 6 per cent., and from that to other per cents. But it is equally applicable for *all* per cents., from 1 to 15 inclusive.

BANKERS' METHOD OF COMPUTING INTEREST. 107

The following table shows the different per cents., with the time that a given number of $ will amount to the same number of cents when placed at interest.

RULE.—*Draw a perpendicular line, cutting off the two right hand figures of $, and you have the interest on the following per cents. :*

    Interest at 4 per cent. for 90 days.
    Interest at 5 per cent. for 72 days.
    Interest at 6 per cent. for 60 days.
    Interest at 7 per cent. for 52 days.
    Interest at 8 per cent. for 45 days.
    Interest at 9 per cent. for 40 days.
    Interest at 10 per cent. for 36 days.
    Interest at 12 per cent. for 30 days.
    Interest at 7-30 per cent. for 50 days.
    Interest at 5-20 per cent. for 70 days.
    Interest at 10-40 per cent. for 35 days.
    Interest at 7½ per cent. for 48 days.
    Interest at 4½ per cent. for 80 days.

NOTE.—The figures on the left of the perpendicular line are dollars, and on the right decimals of $. If the $ are less than 10 prefix a 0.

EXAMPLE 1.—What is the interest of $120 for 15 days at 4 per cent. ?

                                          Days.   Days.
    $120=principal.   15=⅙ of 90.
      6)1 | 20 cts.=int. for 90 days.
          | 20 cts.=int. for 15 days.

EXAMPLE 2.—What is the interest of $132 for 13 days at 7 per cent.?

$132=principal.   13=¼ of 52. (Days)

4)1 | 32 cts.=int. for 52 days.
    | 33 cts.=int. for 13 days.

EXAMPLE 3.—What is the interest of $520 for 9 days at 8 per cent.?

$520=principal.   9=⅕ of 45. (Days)

5)5 | 20 cts.=int. for 45 days.
$1 | 04 cts.=int. for 9 days.

EXAMPLE 4.—What is the interest of $462 for for 64 days at 7½ per cent.?

$462=principal.   64=48+16. (Days. Days. Days.)

3)4 | 62 cts. =int. for 48 days.
  1 | 54 cts. =int. for 16 days.

—————
$6 | 16 cts.=int. for 64 days.

REMARK.—We have now illustrated several examples by the different per cents.; and if the student will study carefully the solution to the above examples, he will in a short time be very rapid in this mode of computing interest.

NOTE.—The preceding mode of computing interest is derived and deduced from the canceling system; as the ingenious student will readily see. It is a short and easy way of finding interest for days when the days are even or aliquot parts; but when they are not multiples, and three or four di-

BANKER'S METHOD OF COMPUTING INTEREST.

visions are ncessary, the canceling system is much more simple and easy. We will here illustrate an example to show the difference: Required the interest of $420 for 49 days at 6 per cent.

```
     Bankers' method.              Canceling meth
2)4 | 20 cts.=int. for 60 days.        4̶2̶0̶—70
 -- | --                          6̶—3̶6̶    6̶
2)2 | 10 cts.=int. for 30 days.          49
5)1 | 05 cts.=int. for 15 days.          70
3)  | 21 cts.=int. for 3 days.         ------
    |  7 cts.=int. for 1 day.         $3.430 Ans.
-------
$3 | 43 cts.=int. for 49 days.
```

The canceling method is much more brief; we simply cancel 6 in 36, and the quotient 6 into 420; there is no divisor left; hence 70×49 gives the interest at *once*.

If the time had been 15 or 20 days, the Bankers' Method would have been equally as short, because 15 and 20 are aliquot parts of 60. The superiority the canceling system has above all others is this: it takes advantage of the *principal* as well as the *time*

For the benefit of the student, and for the convenience of business men, we will investigate this system to its full extent and explain how to take advantage of the *principal* when no advantage can be taken of the *days*. This is one of the most important characteristics of interest, and very often saves much labor. *It should be used when the days are not even or aliquot parts.*

The following table shows the different sums of money (at the different per cents.) that bear 1 cent interest a day; hence the time in days is always the interest in cents; therefore, to find the interest on any of the following notes at the per cent. attached to it in the table, we have the following rule:

RULE.—*Draw a perpendicular line, cutting off the two right hand figures of the days for cents, and you have the interest for the given time.*

Interest of $90 at 4 per cent. for 1 day is 1 cent.
Interest of $72 at 5 per cent. for 1 day is 1 ct.
Interest of $60 at 6 per cent. for 1 day is 1 ct.
Interest of $52 at 7 per cent. for 1 day is 1 ct.
Interest of $45 at 8 per cent. for 1 day is 1 ct.
Interest of $40 at 9 per cent. for 1 day is 1 ct.
Interest of $36 at 10 per ct. for 1 day is 1 ct.
Interest of $30 at 12 per ct. for 1 day is 1 ct.
Interest of $50 at 7.30 per ct. for 1 day is 1 ct.
Interest of $70 at 5.20 per ct. for 1 day is 1 ct.
Interest of $35 at 10.40 per ct. for 1 day is 1 ct.
Interest of $48 at $7\frac{1}{2}$ per ct. for 1 day is 1 ct
Interest of $80 at $4\frac{1}{2}$ per ct. for 1 day is 1 ct
Interest of $24 at 15 per ct. for 1 day is 1 ct

NOTE.—Government bonds are calculated on the base of 365 days to the year. To find the interest on the base of 365 days, ascertain the interest at 360 days, and subtract $\frac{1}{73}$ of the sum obtained.

BANKERS' METHOD OF COMPUTING INTEREST. 111

NOTE.—This table should be committed to memory, as it is very useful when the days are not even or aliquot parts. If the days are less than 10 prefix a 0 before drawing the line.

EXAMPLE 1.—Required the interest of $60 for 117 days at 6 per cent.

117=the days. Here we cut off the two
$1 | 17 cts. *Ans.* right hand figures for cents.

The student should bear in mind that the interest on $60 for 117 days is just the same as the interest on $117 for 60 days.

By looking at the table we see that the interest for $60 at 6 per cent. is 1 cent a day; hence the time in days is the answer in cents. If this note was $120, instead of $60, we would first find the interest for $60, and then double it; if it was $180, we would multiply by 3, etc.

EXAMPLE 2.—Required the interest of $45 for 219 days at 8 per cent.

219=the days. Here we cut off the two
$2 | 19 cts. *Ans.* right hand figures for cents.

The student should bear in mind that the interest on $45 for 219 days is just the same as the interest on $219 for 45 days.

By looking at the table we see that the interest on $45 at 8 per cent. is 1 cent a day; hence the time in days is the answer in cents. If this amt.

was $22.50, instead of $45, we would first get the interest for $45, and then divide by 2; if it was $75, we would add on ⅔; if $60, add on ½, etc.

EXAMPLE 3.—Required the interest of $48 for 115 days at 9 per cent.

115=the days. $48=$40+$8.
5)$1 | 15 cts.=the int. of $40 for 115 days.
  | 23 cts.=the int. of $8 for 115 days.

Ans. $1 | 38 cts.=the int. of $48 for 115 days.

Here we first find the interest of $40, because the days is the interest in cents; then we divide by 5 to find the interest for $8; then by adding both we find the interest for $48, as required.

EXAMPLE. 4—Required the interest of $260 for 104 days at 7 per cent.

$52×5=$260.
 104=the days.
  $1 | 04 cts=the int. of $52 for 104 days.
Ans. $5 | 20 cts. Multiply by 5.

Here we first find the interest of $52, because the days is the interest in cents; then we multiply by 5 to get it for $260. We could have worked this note by the Bankers' Method, just as well, by cutting off two figures in the principal, making $2.60 cts. the interest for 52 days, and then multiply by 2 to get it for 104 days. The student must remember that the interest of $260 for 104 days is just the same as the interest of $104 for 260 days.

BANKER'S METHOD OF COMPUTING INTEREST. 113

Problems Solved by Both Methods.

We will now solve some examples by both methods, to further illustrate this system, and for the purpose of teaching the pupil how to use his judgment. He will then have learned a rule *more valuable than all others*.

EXAMPLE 5.—What is the interest $180 for 75 days at 6 per cent.?

Operation by taking advantage of the $.
75 = the days. $60 × 3 = $180.
$0 | 75 cts. = the int. of $60 for 75 days.
   | 3   Multiply by 3.
___
Ans. $2 | 25 cts. = the int. of $180 for 75 days.

Operation by the Bankers' Method.
$180 = the principal. 60da. + 15da. = 75da.
4)$1 | 80 cts. = the int. for 60 days.
   | 45 cts. = the int. for 15 days.
___
Ans. $2 | 25 cts. = the int. for 75 days.

By the first method we multiplied by 3, because 3 × $60 = $180; by the second method we added on ¼, because 60da. + ⁶⁰⁄₄da. = 75da.

N. B.—When advantage can be taken of both time and principal, if the student wishes to prove his work, he can first work it by the Bankers' Method, and then by taking advantage of the principal, or *vice versa*. And as the two operations are entirely different, if the same result is obtained by each, he may fairly conclude that the work is correct

Lightning Method of Computing Interest.

On all notes that bear $12 per annum, or any aliquot part or multiple of $12.

IF a note bears $12 per annum, it will certainly bear $1 per month; hence the time in months would be the interest in $; and the decimal parts of a month would be the interest in decimal parts of a $; therefore when the note bears $12 per annum we have the following rule:

RULE.—*Reduce the years to months, add in the given months, and place one-third of the days to the right of this number, and you have the interest in dimes.*

EXAMPLE 1.—Required the interest of $200 for 3 years, 7 months, and 12 days, at 6 per cent.

```
   200                           ⅓ of 12 days=4.
     6
   ———                        Yr. Mo. Da.
$12.00=int. for 1 yr.          3   7  12 =43.4mo.
```
Hence 43.4 dimes, or $43.40cts., *Ans.*

We see by inspection that this note bears $12 interest a year; hence the time reduced to months, with one-third of the days to the right, is the interest in dimes. If this note bore $6 a year, instead of $12, we would take one-half of the above interest, if it bore $18, instead of $12, we would add one-half; if it bore $24, instead of $12, we would multiply by 2, etc.

EXAMPLE 2.—Required the interest of $150 for 2 years, 5 months, and 13 days, at 8 per cent.

150
8
———
$12.00=int. for 1 yr.

$\frac{1}{3}$ of 13 days=4$\frac{1}{3}$.

Yr. Mo. Da.
2 5 13=29.4$\frac{1}{3}$mos

Hence 29.4\frac{1}{3}$ dimes, or 29.43\frac{1}{3}$ cts., *Ans.*

We see by inspection that this note bears $12 interest a year; hence the time reduced to months, with one-third of the days placed to the right, gives the interest at once.

EXAMPLE 3.—Required the interest of $160 for 11 years, 11 months, and 11 days, at 7$\frac{1}{2}$ per cent.

160
7$\frac{1}{2}$
———
$12.00=int. for 1 yr.

$\frac{1}{3}$ of 11 days=3$\frac{2}{3}$.

Yr. Mo. Da
11 11 11=143.3$\frac{2}{3}$mos.

Hence 143.3\frac{2}{3}$ dimes, or 143.36\frac{2}{3}$ cts., *Ans.*

When the Interest is more or less than $12 a Year.

RULE.—*First find the interest for the given time on the base of $12 interest a year; then, if the interest on the note is only $6 a year, divide by 2; if*

$24 a year, multiply by 2; if $18 a year, add on one-half, etc.

EXAMPLE 1.—What is the interest of $300 for 4 years, 7 months, and 18 days, at 6 per cent.

⅓ of 18 days=6
300   4yr. 7mo. 18da.=55.6mo
  6

$18.00=int. for 1 year.   2)55.6, int. at $12 a year
$18=1½ times $12.   278

$83.4 *Ans.*

If the interest was $12 a year, $55.60 would be the answer; because 55.6 is the time reduced to months; but it bears $18 a year, or 1½ times 12; hence 1½ times 55.6 gives the interest at once.

EXAMPLE 2.—Required the interest of $150 for 3 years, 9 months, and 27 days, at 4 per cent.

⅓ of 27 days=9.
150   3yr. 9mo. 27da.=45.9mo.
  4   2)45.9, int. at $12 a year.

$6.00=int. for 1 year. $22.95 *Ans.*
$6=½ times $12.

If the interest was $12 a year, $45.90 would be the answer; because 245.9 is the time reduced to months; but it bears $6 a year, or ½ times 12; hence ½ times 45.9 gives the interest at once.

FOR YEARS, MONTHS, AND DAYS.

THE computation of simple interest, where the time consists of years, months, and days, is quite difficult. Taking the aliquot parts for the different portions of time almost invariably involves the calculator in fractions, and, unless he is well versed in vulgar fractions he will not be able to arrive at the correct result. We have three bases by which we compute interest at different rates per cent. and by which we are enabled to entirely avoid the use of fractions. These three bases are each obtained different from the other, and consequently we have three rules for computing interest: one at a base of one per cent., a second at a base of twelve per cent., and a third at a base of thirty-six per cent.

RULE for computing interest at 1 per cent.:

Take one-third of the number of days and annex to the number of months; divide the number thus formed by 12; annex the quotient thus obtained to the number of years, and multiply the principal by this number; if the principal contains cents, point off five decimal places; if not, point off three deci-

mal places; this will give the interest at one per cent. For any other rate per cent., multiply the interest at one per cent. by the required rate per cent.

Remark.—This rule applies to all problems in interest where the days are divisible by 3, and this number, annexed to the number of months, divisible by 12.

EXAMPLE.

Required the interest on $112, at 1 per cent., for 3 years, 3 months and 18 days.

SOLUTION.

Take one-third of the number of days, $\frac{1}{3}$ of 18 =6, annex this number to the months given, 36, divide this number by 12, 36÷12=3, annex this number to the year gives, 33, multiply the principal by 33, $112×33=3.69 6, point off three decimal places, and we have the required interest, $3.69 6.

EXAMPLE.

Required the interest on $125 12, at 7 per cent. for 2 years, 8 months and 12 days.

SOLUTION.

Take one-third of the number of days, $\frac{1}{3}$ of 12=4, annex this number to the number of months we have 84, divide this number by 12,

MERCHANTS' METHOD OF COMPUTING INT. 119

84÷12=7, annex this number to the number of years we have 27, multiply the principal by this number, and point off five decimal places, and you have the interest at one per cent.; multiply this interest by 7, and you have the interest at 7 per cent., the required rate.

$125 12
      27
———————
  3.37824
        7
———————
$23 .64768

EXAMPLE.

Required the interest on $1,023, at 8 per cent., for 1 year, 9 months and 18 days.

SOLUTION.

Take one-third the number of days and annex to the number of months, ⅓ of 18=6, we have 96÷12=8, annex this number to the years we have 18, multiply the principal by this number, and point of three decimal places, which gives the interest at 1 per cent.; multiply the interest at one per cent. by 8, and you have the required interest.

$1023
   18
———————
$18 .414
       8
———————
$147 .312

Remark.—This rule will apply to all problems in interest if one-third of the number of the days be taken decimally and annexed to the number of months, and this number, divided by 12, carried out decimally. But this makes the multiplier very large; hence, to avoid this large number in the multiplier, where the days are divisible by 3 and this number, annexed to the months, is not divisible by 12, we use the following rule, called our base at 12 per cent.:

RULE.—*Reduce the years to months, add in the months, take one-third of the number of days and annex to this number, multiply the principal by the number thus formed; if there are cents in the principal, point off five decimal places; if there are no cents in the principal, point off three decimal places; this gives the interest at 12 per cent. For any other rate per cent., take such part of the base before multiplying as the required rate is a part of* 12.

EXAMPLE.

Required the interest on $123, at 12 per cent for 2 years, 2 months and six days.

SOLUTION.

Reduce the 2 years to months gives us 24 months, add on the 2 months gives us 26 months, take one-third of the days, $\frac{1}{3}$ of $6=2$, annexed to the 26 months gives 262, which constitutes the base; multiply the principal by this base, and you have $32 226 the interest at 12 per cent.

$123
262
———
$32 226

EXAMPLE.

Required the interest on $144, at 6 per cent., for 4 years, 5 months and 12 days.

SOLUTION.

Reduce the 4 years to months gives 48 months, add in the 5 months gives 53 months, take one-third of the days and annex to the number of months, ⅓ of 12=4, annex to the 53 months, 534; this number multiplied into the principal would give the interest at 12 per cent. But we want it at 6 per cent. We will now take such part of either principal or base as 6 is a part of 12; 6 is ½ of 12, therefore we will take ½ of 144=72, one-half of the principal, and multiply it by the base, which will give the interest at 6 per cent.

$$\begin{array}{r} 534 \\ \hline \$38.448 \end{array}$$

EXAMPLE.

Required the interest on $347 25, at 8 per cent., for 2 years, 3 months and 9 days.

SOLUTION.

Reduce the 2 years to months, 24 months, add the 3 months, 27 months, take one-third of the days, ⅓ of 9=3, annex to the months, 273, the base; this, multiplied into the principal, would give the interest at 12 per cent. But we want the interest at 8 per cent; we will take two-thirds of the base before multiplying: ⅔ of 273=182; the principal multiplied by this number gives the interest at 8 per cent.

$$\begin{array}{r} \$347\ 25 \\ 182 \\ \hline \$63.19950 \end{array}$$

Remark.—This base is used where the days are divisible by 3, and the number formed by annexing one-third of the days to the months not divisible by 12. We now come to time in which neither days nor months are divisible. Where such time as this occurs, we use a base at 36 per cent.

RULE.—*Reduce the time to days, by multiplying the years by 12, adding in the months, if any, and multiplying this number by 30, adding in the days, if any; multiply the principal by this number, pointing off 5 decimal places, where cents are given in the principal, and 3 places where no cents are given. This will give the interest at 36 per cent.*

EXAMPLE.

Required the interest on $144, at 36 per cent., for 3 years, 2 months and 2 days.

SOLUTION.

Reduce the time to days gives 1142 days; multiply the principal by this base, and you have the interest at 36 per cent

$144
1142
———
$164.448

EXAMPLE.

Required the interest on $144, at 9 per cent., for 5 years, 7 months and 5 days.

SOLUTION.

Reduce the time to days gives 2,015 days; if we multiply the principal by this base, we would get the interest at 36 per cent.; but we want it at 9 per cent. We can take such part of either principal or base as 9 is a part of 36 before multiplying; 9 is ¼ of 36; we will take ¼ of the principal, it being divisible by 4; ¼ of 144=36, which, multiplied into the base, will give the interest at 9 per cent., by pointing off 3 decimal places.

2015
36
―――――
$72.540

EXAMPLE.

Required the interest on $875 15, at 6 per cent., for 5 years, 7 months and 12 days.

SOLUTION.

Reduce the time to days gives 2022 days; 6 is ⅙ of 36; take one sixth of the base, ⅙ of 2022=337; multiply the principal by this number, point off 5 decimal places, and you have the interest at 6 per cent., the required rate.

$875 15
337
―――――
$294.92555

Remark.—We have now fully explained our method of computing interest at the three different bases. Any and every problem in interest can be solved by one of these three bases. Some problems can be solved easier by one base than another. Where the days are divisible by 3, and their number, annexed to the months, divisible by 12, it is the shortest and best method to use the base at 1 per cent. By using one or the other of these three bases, the student can avoid the use of vulgar fractions. The student must study these three principles carefully, and learn to adopt readily the base best suited to the problem to be solved.

PARTIAL PAYMENTS ON NOTES, BONDS, MORTGAGES

To compute interest on notes, bonds, and mortgages, on which partial payments have been made, two or three rules are given. The following is called the common rule, and applies to cases where the time is short, and payments made within a year of each other. This rule is sanctioned by custom and *common law;* it is true to the principles of simple interest, and requires no special enactment. The other rules are rules of *law*, made to suit such cases as require (either expressed or implied) annual interest to be paid, and of course apply to no business transactions closed within a year.

RULE.—*Compute the interest of the principal sum for the whole time to the day of settlement, and find the amount. Compute the interest on the several payments, from the time each was paid to the day of settlement; add the several payments and the interest on each together, and call the sum the amount of the payments. Subtract the amount of the payments from the amount of the principal, will leave the sum due.*

PARTIAL PAYMENTS. 125

EXAMPLES.

1. A gave his note to B for $10,000; at the end of 4 months, A paid $6,000; and at the expiration of another 4 months, he paid an additional sum of $3,000; how much did he owe B at the close of the year?

By the Common Rule.

```
Principal.................................$10,000
Interest for the whole time.............    600
                                         --------
Amount...................................$10,600
1st payment........$6,000
Interest, 8 months   240
2d payment.........  3,000
Interest, 4 months    60
                   -------
Amount ............$9,300                  9,300
                                         --------
                  Due....................$1 300
```

PROBLEMS IN INTEREST.

There are *four* parts or quantities connected with each operation in interest: these are, the *Principal, Rate per cent., Time, Interest or Amount.*

If any *three* of them are given the *other* may be found.

Principal, interest, and time given, to find the rate per cent.

1. At what rate per cent. must $500 be put on interest to gain $120 in 4 years?

126   ORTON & SADLER'S CALCULATOR.

<table>
<tr><td>Operation.</td><td>By analysis</td></tr>
<tr><td>$500
.01
―――
5.00
4
―――
20.00)120.00(6 per cent., Ans.
120.00
―――</td><td>The interest of $1 for the given time at 1 per cent is 4 cents. $500 will be 500 times as much =500×.04 =$20.00. Then if $20 give 1 per cent., $120 will give $\frac{120}{20}$ =6 per cent.</td></tr>
</table>

RULE.—*Divide the given interest by the interest of the given sum at 1 per cent. for the given time, and the quotient will be the rate per cent. required*

Principal, interest, and rate per cent. given, to find the time.

2. How long must $500 be on interest at 6 per cent. to gain $120?

<table>
<tr><td>Operation</td><td>By analysis.</td></tr>
<tr><td>$500
.06
―――
30.00)120.00(4 years, Ans.
120.00
―――</td><td>We find the interest of $1.00 at the given rate for 1 year is 6 cents $500, will therefore be 500 times as</td></tr>
</table>

much=500×.06=$30.00. Now, if it take 1 year to gain $30, it will require $\frac{120}{30}$ to gain $120=4 years. *Ans.*

Rule.—*Divide the given interest by the interest of the principal for 1 year, and the quotient is the time.*

Given the amount, time, and rate per cent., to find the *principal.*

Rule.—*Divide the given amount by the amount of $1, at the given rate per cent., for the given time.*

Remark.—This rule is deduced from the fact that the amount of different principals for the same time and at the same rate per cent., are to each other as those principals.

BANK DISCOUNT.

Bank Discount is the sum paid to a bank for the payment of a note before it becomes due.

The amount named in a note is called the *face* of the note. The *discount* is the interest on the face of the note for 3 days more than the time specified, and is paid in advance. These 3 days are called *days of grace*, as the borrower is not obliged to make payment until their expiration. Hence, to compute bank discount, we have the following

Rule.—*Find the interest on the face of the note for 3 days more than the* TIME *specified; this will be the discount. From the face of the note deduct the discount, and the remainder will be the* PRESENT VALUE *of the note.*

DISCOUNT, OR COUNTING BACK.

The object of discount is to show us what allowance should be made when any sum of money is paid before it becomes due.

The *present worth* of any sum is the principal that must be put at interest to amount to that sum in the given time. That is, $100 is the *present worth* of $106 due one year hence; because $100 at 6 per cent. will amount to $106; and $6 is the *discount*.

1. What is the present worth of $12.72 due one year hence?

```
    First method.              Second method.
       $12.72                      $
          100                  1.06)12.72($12 Ans.
      ———                          10.6
106)1272.00($12 Ans.              ———
       106                          2.12
      ———                           2.12
        212                       ———
        212
      ———
```

As $100 will amount to $106 in one year at 6 per cent., it is evident that if $\frac{100}{106}$ of any sum be taken, it will be its present worth for one year, and that $\frac{6}{106}$ will be the discount. And as $1 is the present worth of $1.06 due one year hence, it is evident that the present worth of $12.72 must be equal to the number of times $12.72 will contain $1.06.

RULE.—*Divide the given sum by the amount of $1 for the given rate and time, and the quotient will be the present worth. If the present worth be subtracted from the given sum, the remainder will be the discount.*

EQUATION OF PAYMENTS

EQUATION OF PAYMENTS is the process of finding the equalized or average time for the payment of several sums due at different times, without loss to either party.

To find the average or mean time of payment, when the several sums have the same date.

RULE.—*Multiply each payment by the time that must elapse before it becomes due; then divide the sum of these products by the sum of the payments, and the quotient will be the averaged time required.*

NOTE.—When a payment is to be made down, it has no product, but it must be added with the other payments in finding the average time.

EXAMPLE 1.—I purchased goods to the amount of $1200; $300 of which I am to pay in 4 months, $400 in 5 months, and $500 in 8 months. How long a credit ought I to receive, if I pay the whole sum at once? *Ans.* 6 months.

Mo.   Mo.
4×300=1200   { A credit on $300 for 4 months is the same as the credit on $1 for 1200 months.
5×400=2000   { A credit on $400 for 5 months is the same as the credit on $1 for 2000 months.
8×500=4000   { A credit on $500 for 8 months is the same as the credit on $1 for 4000 months.

1200) 7200 (6 mo.   Therefore, I should have the same credit as a credit on $1 for 7200 months, and on $1200, the whole sum, one-twelfth hundredth part of 7200 months, which is 6 months.
7200

This rule is the one usually adopted by merchants, although not strictly correct, still, it is sufficiently accurate for all practical purposes.

To find the average or mean time of payment, when the several sums have different dates.

EXAMPLE 1.—Purchased of James Brown, at sundry times, and on various terms of credit, as by the statement annexed. When is the *medium* time of payment?

Jan.   1, a bill am'ting to $360, on 3 months' credit.
Jan. 15, do.   do.   186, on 4 months' credit.
March 1, do.   do.   450, on 4 months' credit.
May 15, do.   do.   300, on 3 months' credit
June 20, do.   do.   500, on 5 months' credit

Ans. July 24th, or in 115 da.

Due April 1, $360
    May 15, 186× 44=   8184
    July 1, 450× 91= 40950
    Aug. 15, 300×136= 40800
    Nov. 20, 500×233= 116500
              ―――――   ―――――――
              1796+ into)206434(114$\frac{230}{1796}$ days.

EQUATION OF PAYMENTS.

We first find the time when each of the bills will become due. Then, since it will shorten the operation and bring the same result, *we take the time when the first bill becomes due*, instead of its *date*, for the *period* from which to compute the average time. Now, since April 1 is the period from which the average time is computed, no time will be reckoned on the first bill, but the time for the payment of the second bill extends 44 days beyond April 1, and we multiply it by 44.

Proceeding in the same manner with the remaining bills, we find the average time of payment to 114 days and a fraction, from April 1, or on the 24th of July.

RULE.—*Find the time when each of the sums becomes due, and multiply each sum by the number of days from the time of the earliest payment to the payment of each sum respectively. Then proceed as in the last rule, and the quotient will be the average time required, in days, from the earliest payment.*

NOTE.—Nearly the same result may be obtained by reckoning the time in months.

In mercantile transactions it is customary to give a credit of from 3 to 9 months, on bills of sale Merchants in settling such accounts, as consist of various items of debit and credit for different times, generally employ the following:

RULE.—*Place on the debtor or credit side, such a sum, (which may be called* MERCHANDISE BALANCE,) *as will balance the account.*

Multiply the number of dollars in each entry by the number of days from the time the entry was made to the time of settlement; and the Merchandise balance by the number of days for which credit was given. Then multiply the difference between the sum of the debit, and the sum of the credit products, by the interest of $1 for 1 day; this product will be the INTEREST BALANCE.

When the sum of the debit products exceed the sum of the credit products, the interest balance is in favor of the debit side; but when the sum of the credit products exceed the sum of the debit products, it is in favor of the credit side. Now to the merchandise balance add the interest balance, or subtract it, as the case may require, and you obtain the CASH BALANCE.

A has with B the following account:

1849.		Dr.	1849.		Cr.
Jan. 2.	To merchandise,	$200	Feb. 20.	By merchandise,	$100
April 20.	" "	400	May. 10.	" "	300

If interest is estimated at 7 per cent., and a credit of 60 days is allowed on the different sums, what is the cash balance August 20, 1849?

Ans. 206.54.

EXPLANATION.— Without interest the cash balance would be $200

EQUATION OF PAYMENTS. 133

If no credit had been given, the debits should be increased by the interest of $200 for 230 days, at 7 per cent.; and the interest of $400 for 122 days, at 7 per cent. The credits should be increased by the interest of $100 for 181 days, at 7 per cent., and the interest of $300 for 102 days, at 7 per cent.

Since a credit of 60 days is given on all sums, it is evident by the above calculation, that we should increase the debits by the interest of the sum of the debits, $600, for 60 days more than justice requires. Also, that we should increase the credits by the interest of the sum of the credits, $400, for 60 days more than we should do.

Now, instead of deducting these items of interest from the *amount* of debit and credit interests, it is plain that it will be more convenient and equally just, to diminish the debit interest of the *merchandise balance* for 60 days, which can be most readily accomplished by adding the interest on the merchandise balance for 60 days, to the credit items of interest.

From which we discover that the *interest balance* is equal to the difference between the sum of the debit interests, and the sum of the credit interests increased by the interest of the merchandise balance for the time for which credit was given.

Operation.

```
     DEBITS.                           CREDITS.
  $      Days.                      $      Days.
 200 × 230 = 46000               100 × 181 = 18100
 400 × 122 = 48800               300 × 102 = 30600
 ─────────────────   Balance, 200 ×  60 = 12000
           94800                 ─────────────────
           60700                           60700
```

$$\frac{0.07}{365} \times 34100 = \$6.54 \text{ } \textit{Interest balance, yearly.}$$

Therefore, the foregoing account becomes balanced as follows:

```
1849.                     Dr. | 1849.                        Cr.
Jan.  2. To Merchandise $200.00 | Feb. 20. By Merchandise, $100.00
April 20.  "      "      400.00 | May. 10.  "       "       300.00
Aug. 20.   " balance of int. 6.54 | Aug. 20. " balance,       206.54
                        ────────                         ────────
                         $606.54                          $606.54
Aug. 20. " Cash balance, $206.54 |
```

NOTE.—It is customary in practice, when the number of cents in any of the entries are less than 50, to omit them, and to add $1 when they are 50 or more.

LIGHTNING METHOD
OF
AVERAGING ACCOUNTS.

From "PACKARD'S KEY TO COMPLETE COURSE." With kind permission of the author, S. S. Packard, President of Packard's New York Business College.

The matter of averaging accounts, or which is usually styled in the arithmetics, "Compound Average," is of such immediate and vital importance to the accountant, that we submit here a *short method* which, from its mechanical advantages and other considerations, has been well-received by those who have had occasion to use it. For reasons that will be apparent, it is not theoretically exact, but the discrepancies are *so small* and unimportant that they are scarcely worthy a thought, practically speaking, and besides as a fraction of a day in the results of average cannot be considered, the slight incidental variations in this method have no real bearing on the result.

This method has been in use for the past fifteen or twenty years, and its main features have been published by different authors; and yet certain itinerant professors are in the habit of claiming its originality, and offering, under pledge of secrecy, and for a substantial consideration in hand paid, to disclose its wonderful properties.

It is neither more nor less than a convenient mechanical arrangement, whereby the principles of average are effectively and compactly enforced.

We will illustrate its workings from the materials contained in the following example:

EXAMPLE.

When is the balance of the following account due?

S. S. PACKARD.

1871	Debits	1871	Credits
May 12	750	June 10	500
" 30	117	" 30	300
June 12	340		
July 1	150	Due by average, April 26.	

AVERAGING ACCOUNTS.

ILLUSTRATION.

NOTE.—The assumed date is fixed on the 31st of December, preceding the earliest date in the account. This is for convenience sake, and to preserve constant uniformity. Some prefer the last day of the month preceding the first item. Interest is reckoned on each amount from December 31 to the date due at 12% per annum, or 1% per month. This rate of percentage or interest is also arbitrarily fixed, as being the most convenient for use.

```
   Debits.              PROCESS.              Credits.
May 12...$750  { Int. 4 mos., 30 00    June 10...$500  { Int. 5 mos., 25 00
               { "   12 days,  3 00                    { "  10 days,  1 67
 "   30....117 { "    5 mos.,  5 85     "   30....300  { "   6 mos., 18 00
                {"    5 mos., 17 00
June 12....340 {"   12 days,  1 36    Total Cr. $ 800   Tot. Cr.Int. 44 67
                {"    6 mos.,  9 00
July 1.....150 {"    1 day,     05

Total Dr. )         ( Total Dr.
  of $    } 1357    { of Int.  } 66 26
Total Cr. )         ( Total Cr.
  of $    }  800    { of Int.  } 44 67
Bal. of $    557  Bal. of Int. 21 59(3 months.
                               16 71
                                4 88
                                  30
                              14640(26 + days.
                               1114
                               3500
                               3342
                                158
```

Ans.—Balance of $ due in 3 mos., 26 days, from Dec. 31, or April 26.

We regret that our limited space will not permit a more detailed exhibit of the method, either as to its philosophy or its facts. The above working must carry its own suggestions. The theory is, that if a settlement were made as on the preceding 31st of December (the assumed day of settlement), the *debit side* of the account, as shown, would be entitled to $66.26 discount, and the *credit side* to $44.67, making a balance of $21.59 in favor of the debit side. As the balance of account is also in favor of the debit side, it is only necessary to know how long it would take the balance of account to produce the balance of interest (at the rate of 1% a month, or 12% per annum*); to know the *time—reckoned forward—*when the balance of account falls due. Now, as the rate named (1% a month), the interest for one month can be had by merely cutting off two figures from the right of dollars, we have the balance of account thus divided ($5.57), a ready divisor of the balance of interest ($21.59), the quotient being the number of months and parts of a month it will take the *balance of account* to produce the *balance of interest*. This time reckoned forward from the assumed focal date, will get the average date of payment.

Thus, 3 months, 26 days from December 31 will bring the average date as stated—April 26th.

---

*FORMULA.

For calculating Interest at 12% per annum.

Time—Months and days.

The Principal Dollars only. { Multiplied by the number of months. } = Interest expressed in cents.
{ Multiplied by ⅓ the number of days. } = Interest expressed in mills.

⅓ the Principal Dollars only. { Multiplied by any number of days. } = Interest expressed in mills.

NOTE.—In the above application, when the principal contains cents, point off two additional decimal places.

PARTNERSHIPS

Is the association of two or more persons under a copartnership name, for the purpose of transacting business for their mutual profit, with certain agreements regarding the investment of capital and the division of gains or losses between them.

The Partners are the persons associated together in business.

The Capital or Stock is the cash or property invested.

The Resources of a firm or copartnership is the property owned by them, including money and claims due from others.

The Liabilities are the debts owed by the firm or claims against the copartnership.

Net Capital is the excess of resources over the liabilities.

Net Insolvency is the excess of liabilities over the resources.

Net Gain is the excess of gains over the losses and expenses, and is shown by taking the difference between the net capital at commencing business from that shown at closing the books, which is termed *Present Worth*.

Net Loss is the excess of losses over the gains or profits.

DIVISION OF GAINS OR LOSSES

Adjusted between parties according to capital invested.

In the adjustment of gains or losses between copartners on the basis of capital invested, we have three methods, *i. e.*, *Percentage*, *Fractions*, and *Proportion*, each producing in the aggregate the same result.

FIRST METHOD—BY PERCENTAGE.

RULE I.—*Ascertain the per cent. of gain or loss* (on capital invested) *by dividing the net gain or loss by the net capital. For each partner's share of gain or loss, multiply his net investment by the percentage thus obtained.**

SECOND METHOD—BY FRACTIONS.

NOTE.—As there will be as many fractional parts in the adjustment as there are partners, we produce the following:

RULE II.—*Take the net investment of each partner for the numerator of a fraction, and the net capital for its denominator; and for each partner's share take his respective fractional part of the entire gain or loss.**

---

* Each partner's share of **gain** or **loss** added together equals the total gain or loss.

PARTNERSHIP.

THIRD METHOD—OR PROPORTION.

NOTE.—This method we consider the most simple and practical, and advise its use in preference to either of the above methods, except in simple cases of adjustment, when the division can be shown by small fractions, thus, $\frac{1}{2}+\frac{1}{3}+\frac{1}{6}=$ the total interest of partners.

RULE III.—*State by proportion, as the total capital is to each partner's investment, so is the net gain or loss to each partner's respective share of same.**

Proper form of proportional statement.

Total capital of partners. : Each partner's share. :: Net gain or loss. : Each partner's gain or loss.

Working.—Multiply each partner's share by the net gain or loss, divide that product by the total capital, and the quotient will be each partner's respective share of gain or loss.

EXAMPLE.

E. Burne, W. H. Devon, and J. K. Hopper are copartners, the gains or losses arising from business to be divided between them in proportion to average investment. The investments are as follows: Mr. B. $5500; Mr. D. $6750; Mr. H. $3250. Upon closing the books the net gain is found to be $3850—what is each partner's respective share? *Ans.* Mr. B. $1366.13; Mr. D. $1676.61; Mr. H. $807.26.

---

* Each partner's share of gain or loss added together equals the total gain or loss.

Solution.

B. investment, 5500 ⎫
D.   "   6750 ⎬ Total am't, 15,000. : ⎧ B. invest., 5500 ⎫
H.   "   3250 ⎭   ⎨ D.   "   6750 ⎬ :: { Net gain 3850 }
                                                  ⎩ H.   "   3250 ⎭

$5500 \times 3850 = 21175000 \div 15500 = 1366.13$ B's share
$6750 \times 3850 = 25987500 \div 15500 = 1676.61$ D's   "
$3250 \times 3850 = 12512000 \div 15500 = \overline{807.26}$ H's   "
               Total gain, $3850.00—or Proof.

DIVISION OF GAINS OR LOSSES

Between partners, according to investment, when the capital is furnished at different dates.

RULE.—*Multiply each partner's investment by the time employed. The product thus obtained equals the average investment for the average time, and the sum of the products the total average capital for the average time.*

NOTE.—After ascertaining the average capital and investment by the above rule, to secure each partner's respective share of the gain or loss, proceed according to either rule on pages 140, 141.

EXAMPLE.

A and B are partners, gains or losses to be divided according to average investment.

A puts in Jan. 1, $5000 | B puts in Jan. 1, $2000
"   "   Feb. 1, 1000 | "   "   Apr. 1. 3000
"   "   Sep. 1, 2000 | "   "   July 1, 1000

January 1, one year from date of first investment, the books are closed, and the *net gain* ascertained to be $2720—what is each partner's share? *Ans.* A's $1580; B's $1140.

PARTNERSHIP. 143

Process of Solution.

```
                          Invest-  Time,  Average  Average
                          ment.    mos.   Capital. Time.
A invested Jan. 1 to Jan. 1, 5000 ×12= 60000 —1 mo.
 "     "       Feb. 1 "   "  1000 ×11= 11000 —1 "
 "     "       Sep. 1 "   "  2000 × 4=  8000 —1 "
A's aver. investm't for the aver. time, 79000 —1 "

B invested Jan. 1 to Jan. 1, 2000 ×12= 24000 —1 "
 "     "       Apr. 1 "   "  3000 × 9= 27000 —1 "
 "     "       July 1 "   "  1000 × 6=  6000 —1 "
B's aver. investm't for the aver. time, 57000 —1 "
A's   "       "       "    "    "      79000 —1 "
A & B's average investment=Total
     capital for average time,         136000 —1 "
```

Statement per Rule, page 141.

```
Total aver.  Each partner's  Total   Each partner's
  capital.   aver. capital.   gain.  share of gain.
 136000  :    79000    : :    2720 = 1580 —A's
 136000  :    57000    : :    2720 = 1140 —B's
```

Working.

79000 × 2720 = 214880000 ÷ 136000 = $1580 A's share.
57000 × 2720 = 155040000 ÷ 136000 = 1140 B's "
                            Total gain, $2720 Proof.

GOLD TO CURRENCY

WHEN GOLD COMMANDS A PREMIUM.

RULE.—*Multiply the amount by 100, increased by the rate per cent. of premium, and point off four decimal places, if there are cents in the amount; if no cents occur, point off but two places.*

Example.— What amount in currency will $666.66⅔ gold purchase, when at a premium of 50 per cent.?

Process.— 100 + 50 = 150.
$666.66⅔ × 150 = $1000 currency.

CURRENCY TO GOLD.

WHEN GOLD COMMANDS A PREMIUM.

RULE.—*Divide the amount by 100 increased by the rate per cent. of premium, and point off two less than the number of places in the dividend.*

Example.—What amount in gold will $1000 currency purchase, gold commanding a premium of 50 per cent.?

Process.— 100 + 50 = 150.
$1000 ÷ 150 = $666.66⅔ gold

```
150 ) 1000.0,0,0,0 ( 666.66⅔
      900
      ----
      1000
       900
       ----
       1000
        900
        ----
        1000
         900
         ----
         100
```

TO ASCERTAIN THE GOLD VALUE OF CURRENCY WHEN AT A DISCOUNT.

Rule.—*Multiply the amount by the rate per cent. of discount; from which subtract the result, and you will have the sum required.*

Example.—When currency is at a discount of 33⅓ per cent., what sum in gold will $1000 purchase?

Process.— $1000 × 33⅓ = discount.
1000 — discount = sum.

```
    1000              1000
    33⅓             333.33
  -------           -------
  $333.33           $666.67
```

$1000 in currency, at 33⅓ per cent. discount, will purchase $666.67 gold.

Note.—When gold commands a premium of 50 per cent., U. S. currency is at 33⅓ per cent. discount.

MATURITY OF COMMERCIAL PAPER.

To ASCERTAIN the Maturity of Commercial Paper payable in days after date.—Set down the days in full; from which take the number of days remaining in the month from date of the paper; and from this result, continue to subtract the number of days contained in the months following, until the remainder is less than 30 (except in case of February), to which add *three days of grace*, and you will have the date of maturity.

Example.—Note, dated March 10, 1874. Payable ninety days after date. Find the date of maturity.

```
Time, days............................. 90
Number of days from March 10 to April 1.. 21
                                          ──
                                          69
                April................. 30
                                          ──
                                          39
                May................... 31
                                          ──
                                           8
Add days of grace................... 3
Date of maturity..........June 11
```

PROOF.

```
Days remaining in March.. 21
  "        "      " April .. 30
  "        "      " May .... 31
  "        "      " June ... 11
                             ──
                             93 ds., including 3 ds. gr.
```

NOTE.—When a Note or Draft falls due on Sunday or any legal holiday authorized by the State or General Government, it must be paid on the day previous. Should a legal holiday occur on Monday, all paper maturing on that day must be paid on the Saturday previous.*

When the time of a Note or Bill is given in days, the days of date and maturity are *counted but one.*

Commercial Paper falling due on the 30th or 31st of any month which contains only 28, 29, or 30 days, becomes due on the last day of the month, hence is legally due on the 3d of the month following.

*Unless otherwise legalized as in the State of New York.

STERLING EXCHANGE.

STERLING EXCHANGE consists of Bills, principally issued by Banks and Bankers upon their correspondents in different countries, to be used in settlement of balances, and conducting of business, without the necessity and risk of special gold remittances.

PREMIUM AND DISCOUNT,

On Bills of Exchange, is regulated by the supply and demand, the same as any marketable commodity.

They command a premium when the balance of trade is against the country where issued, and are subject to discount when in its favor.

BILLS OF EXCHANGE

Are drawn in sterling money, the denominations of which are shown as follows:

4 farthings equal 1 penny—d. | 2 shillings equal 1 florin—fl.
12 pence equal 1 shilling—s. | 20 shillings " 1 pound—£.

The reduction or value of English money in U. S. Gold coin is now based upon the U. S. Standard Value of $4.8665 to the pound, in accordance with the new Act of Congress, which went into effect January 1, 1874.

This is equal to 9½% premium on the old par value of $4.44⁴⁄₉ to the £. The following invaluable tables, for the use of bankers and business men, will save much time and labor in exchange calculations. See pages 152–153.

We also present the old method of calculations.

FOREIGN EXCHANGE QUOTATIONS

Are commonly based upon the nominal value of the £ sterling, which is $4.44⁴⁄₉.

The true value of the £ sterling is...$4.8675
Nominal " " " " ... 4.4444—

Difference........................... .4230

Amounting to nearly 9½ of the nominal. Therefore, when the £ is quoted at 109½, it is really at par.

STERLING EXCHANGE TO U. S. MONEY.

£—s—d—*reduced to Dollars and Cents.*

RULE.—*Reduce the pounds, shillings, and pence to*

STERLING EXCHANGE. 149

sixpences, and divide by 9, the quotient will be the result in dollars and cents; and for every penny exceeding six in the given number of pence, add two cents additional.

NOTE.—Whenever Exchange commands a premium, the per cent. of premium must be added to ascertain the true value; when at a discount, the per cent. must be deducted.

EXPLANATION OF RULE.—*The nominal value of the pound sterling in U. S. money being (4.44\frac{4}{9}$) 4\frac{4}{9}$ = $\frac{40}{9}$ dollars; and in English money, 20 shillings, or 40 sixpences. Therefore—*

$$\begin{array}{rcl} 40 \text{ sixpences} &=& \tfrac{40}{9} \text{ dollars.} \\ 1 \text{ ``} &=& \tfrac{1}{9} \text{ \phantom{dollars}``} \\ 9 \text{ ``} &=& 1 \text{ \phantom{dollars}``} \end{array}$$

Hence, the division by 9 according to the Rule.

PROCESS OF REDUCTION.

£—s—d—to Sixpences.

RULE.—*Multiply the pounds by 40, and the shillings by 2; to the product add 1 whenever the pence equal or are in excess of 6.*

ILLUSTRATION.

Example I.—Reduce £130 9s. 8d. to sixpences.

$$\begin{array}{rrcrl} £ & 130 \times 40 &=& 5200 & \text{sixpences.} \\ s. & 9 \times 2 &=& 18 & \text{``} \\ d. & 8 \div 6 &=& 1 & \text{``} \\ & & & \overline{5219} & \text{``} \end{array}$$

Example II.—What is the par value of £112 9s. 11d., in U. S. gold?

Process.— £ 112 × 40 = 4480 sixpences.
s.   9 × 2 =   18   "
d. 11 ÷ 6 =   1   "
         9) 4499   "
             499.89
Excess of pence 5 × 2 =   .10
             $499.99 U. S. gold.

EXCHANGE, WHEN AT A PREMIUM, OR ABOVE PAR.

Example III.—What is the value, in U. S. gold, of £162 10s. 9d., premium 10½ per cent.?

Process.— £ 162 × 40 = 6480 sixpences.
s. 10 × 2 =   20   "
d.   9 ÷ 6 =   1   "
         9) 6501   "
             $722.33
Excess of pence 3×2 =   6
             $722.39
Premium 10½ per cent. 1.105
             361195
             72239
             72239
             $798.24095
             $798.24 U. S. gold.

STERLING EXCHANGE TO U. S. CURRENCY,

With Exchange and Gold at a premium, or above par.

Example IV.—What will be the cost of a Bill of Exchange on London for £24 14s.?

STERLING EXCHANGE.

Rate of Exchange............ 1.07¾
Gold...................... 1.08¾

Process. — £ 24 × 40 = 960
   s. 14 × 2 = 28
      9) 988
       109.777 gold = 109.78

Exchange quoted 1.07¾............. 1.0775
            54890
            76846
            76846
            10978

Cost in gold.................$118.28
Gold quoted 1.08¾............ 1.0875
            59140
            82796
            94624
            11828

Cost in U. S. currency $128.63

TABLE OF STERLING EXCHANGE,

Showing the value of 1 pound to 1000 in dollars and cents—calculated at the par value of $4.444 to £ sterling.

£	$ Cts.	£	$ Cts.	Sh.	$ Cts.
1	4.444	20	88.889	1	.222
2	8.889	25	111.111	2	.444
3	13.333	50	222.222	3	.667
4	17.778	75	333.353	4	.889
5	22.222	100	444.444	5	1.111
6	26.667	200	888.889	6	1.333
7	31.111	250	1111.111	7	1.778
8	35.556	500	2222.222	10	2.222
9	40.000	750	3333.333	15	3.333
10	44.444	1000	4444.444	20	4.444

TABLE FOR THE
ENGLISH MONEY TO

1 POUND = $4.8665

	0	1	2	3	4
1	4.8665	53.5315	58.3980	63.2645	68.1310
2	9.7330	102.1965	107.0630	111.9295	116.7960
3	14.5995	150.8615	155.7280	160.5945	165.4610
4	19.4660	199.5265	204.3930	209.2595	214.1260
5	24.3325	248.1915	253.0580	257.9245	262.7910
6	29.1990	296.8565	301.7230	306.5895	311.4560
7	34.0655	345.5215	350.3880	355.2545	360.1210
8	38.9320	394.1865	399.0530	403.9195	408.7860
9	43.7985	442.8515	447.7180	452.5845	457.4510

1s. EQUALS 24 133-400 CTS.   1d. EQUALS 2 133-4800 CTS.

	0	1	2	3	4	5	6	7	8	9
0	.....	.2433	.4866	.7299	.9733	1.2166	1.4599	1.7032	1.9466	2.1899
1	.0202	.2636	.5069	.7502	.9935	1.2369	1.4802	1.7235	1.9668	2.2102
2	.0405	.2838	.5272	.7705	1.0138	1.2571	1.5005	1.7438	1.9871	2.2304
3	.0608	.3041	.5474	.7908	1.0341	1.2774	1.5207	1.7641	2.0074	2.2507
4	.0811	.3244	.5677	.8110	1.0544	1.2977	1.5410	1.7843	2.0277	2.2710
5	.1013	.3447	.5880	.8313	1.0746	1.3180	1.5613	1.8046	2.0479	2.2913
6	.1216	.3649	.6083	.8516	1.0949	1.3382	1.5816	1.8249	2.0682	2.3115
7	.1419	.3852	.6285	.8719	1.1152	1.3585	1.6018	1.8452	2.0885	2.3318
8	.1622	.4055	.6488	.8921	1.1355	1.3788	1.6221	1.8654	2.1088	2.3521
9	.1824	.4258	.6691	.9124	1.1557	1.3991	1.6424	1.8857	2.1290	2.3724
10	.2027	.4460	.6894	.9327	1.1760	1.4193	1.6627	1.9060	2.1493	2.3926
11	.2230	.4663	.7096	.9530	1.1963	1.4396	1.6829	1.9263	2.1696	2.4129

NOTE.—To find the value of any number of pounds represented by one figure, find the figure in the left-hand margin of the upper table, and its value will appear in the column adjoining, opposite that figure. To find the value when expressed by two figures, look for the *tens* in the left-hand column, and for the *units* in the top margin, and the value will be shown in the place where the two columns meet; thus, the value of £39 is $189.7935. To find the value of £252, look for 25 as before, and move the decimal point one place to the right, and it shows $1216.625; then add £2 as already shown, $9.7330, and it gives the sum of $1226.358.

REDUCTION OF
U. S. GOLD COIN.
— IN U. S. GOLD.

5	6	7	8	9	
72.9975	77.8640	82.7305	87.5970	92.4635	1
121.6625	126.5290	131.3955	136.2620	141.1285	2
170.3275	175.1940	180.0605	184.9270	189.7935	3
218.9925	223.8590	228.7255	233.5920	238.4585	4
267.6575	272.5240	277.3905	282.2570	287.1235	5
316.3225	321.1890	326.0555	330.9220	335.7885	6
364.9875	369.8540	374.7205	379.5870	384.4535	7
413.6525	418.5190	423.3855	428.2520	433.1185	8
462.3175	467.1840	472.0505	476.9170	481.7835	9

1 *far.* Equals $ .00506.

10	11	12	13	14	15	16	17	18	19	
2.4332	2.6765	2.9190	3.1632	3.4065	3.6498	3.8932	4.1365	4.3798	4.6231	0
2.4535	2.6968	2.9401	3.1835	3.4268	3.6701	3.9134	4.1568	4.4001	4.6434	1
2.4738	2.7171	2.9604	3.2037	3.4471	3.6904	3.9337	4.1770	4.4204	4.6637	2
2.4940	2.7374	2.9807	3.2240	3.4673	3.7107	3.9540	4.1973	4.4406	4.6840	3
2.5143	2.7576	3.0010	3.2443	3.4876	3.7309	3.9743	4.2176	4.4609	4.7042	4
2.5346	2.7779	3.0212	3.2646	3.5079	3.7512	3.9945	4.2379	4.4812	4.7245	5
2.5549	2.7982	3.0415	3.2848	3.5282	3.7715	4.0148	4.2581	4.5015	4.7448	6
2.5751	2.8185	3.0618	3.3051	3.5484	3.7918	4.0351	4.2784	4.5217	4.7651	7
2.5954	2.8387	3.0821	3.3254	3.5687	3.8120	4.0554	4.2987	4.5420	4.7853	8
2.6157	2.8590	3.1023	3.3457	3.5890	3.8323	4.0756	4.3190	4.5623	4.8056	9
2.6360	2.8793	3.1226	3.3659	3.6093	3.8526	4.0959	4.3392	4.5826	4.8259	10
2.6562	2.8996	3.1429	3.3862	3.6295	3.8729	4.1162	4.3595	4.6028	4.8462	11

The lower table shows the value of every combination of shillings and pence less than £1; the upper margin representing the shillings, and the left-hand margin the pence. Thus, to find the value of 13 shillings and 6 pence, follow the column 13 downward until it meets the left-hand column opposite 6, and it shows $3.28. By this method, any number of pounds, shillings, and pence can be reduced to United States gold quickly and accurately.

To ascertain the value of £, s., and d. in U. S. currency, multiply the amount after the reduction, per the above tables, by $1.00, added to the current rate of premium on gold.

MARKING GOODS.

In marking goods it is the custom with most business houses to use a private mark to denote the cost and selling price of the different articles.

Various devices are used to render the *cost* and *selling price* marks from being understood by any except to those employed in the establishment, whose duty it is to exhibit and sell goods.

Any word or phrase containing ten different letters or characters is selected, and used to represent the nine digits and cipher.

To illustrate we will take the following:

GOD HELP US. X.
1 2 3   4 5 6 7   8 9   0

It is required to write a tag or article showing the *cost* and *selling price*.

Assuming the cost to be $2.75, and selling price $3.68, the proper mark will be *ope-dlu*.

An extra letter called a *repcater* is used to prevent the repetition of a letter or figure.

Thus, instead of writing 755 according to the *key word*, which would be *hee*, the repeater K or any other letter not shown in the key could be used, which would make 755 read *hek*. The object of the repeater is to prevent any cue being given to the private mark. Fractions may be used with the letters or characters same as figures. It is usual to write the cost mark above the line, and the selling price below, or *vice versa*, thus, $\frac{ope}{diu}$. The rate of discount from the long price, in favor of the wholesale purchaser, is sometimes written as shown in the above illustration.

Instead of letters, characters or signs may be used. The following words or phrases are adapted for key words:

BLACKHORSE.	IMPORTANCE.
CASH PROFIT.	NOW BE SHARP.

ASKING PRICE AND DISCOUNTS.

It is customary among jobbers and wholesale dealers, in selling their merchandise or wares, to allow certain discounts from the trade or asking price. In this connection we desire to call attention to the fact that losses may arise when supposed profits are being made.

TO ILLUSTRATE.

Suppose our asking price is 30% above the

cost, and we offer a wholesale discount of 15%, our profit is not 15% but 10½%; for the reason the discount is *not* calculated on the *cost* but upon the *asking* price, which includes the original cost and the per cent. of profit added.

EXAMPLE.—Sold an invoice of goods which cost us $100; our asking price being 30% advance or $130, from which we allowed a wholesale discount of 15%—what was the real per cent. of gain? *Ans.* 10½%.

PROCESS.

100 % =	Cost,	or	$ 100.
30 % =	Amount added for profit,	"	30.
130 % =	Asking price,	"	130.
15% discount on $130=19½%.			
19½% =	Wholesale discount,	or	19.50
110½% =	Wholesale net price,	"	110.50
100 % =	Cost,	"	100.
10½% =	Net profit,	"	$10.50 *Ans.*

EXAMPLE II.—Sold an invoice of hats costing $100, they having been marked to sell at a profit of 40%, but in consideration of cash payment we allow a special discount amounting to 30% from our asking price—what is *our net result?* *Ans.* A loss of 2%.

PROCESS.

100% =	Cost,	or	$100.
40% =	Amount added for profit,	"	40.
140% =	Asking price,	"	140.
42% =	30% from asking price, $140,	"	42.
98% =	Amount of cash sales,	"	98.
100% =	Cost,	"	100.
98% =	Amount received,	"	98.
2% =	Net loss.	"	$2.00

TO MARK GOODS

That a discount may be allowed from the asking price and have them net the desired profit.

RULE.—*Assume 100 to be the new or fancy price, from which deduct the proposed discount; then ascertain how many times the amount is contained in the net or original price asked.*

NOTE.—The amount will generally represent hundredths (100), but to avoid fractions it may be extended to thousandths (1000), or tens of thousands (10,000), in which case it will only be necessary to add to the original asking price two, three, or four ciphers, as the case may require.

EXAMPLE I.—A manufacturer wishes to offer the trade a line of goods, which he desires to net $17.00 per dozen—what must be his *asking price* to enable him to allow a *discount* of 15%? *Ans.* $20 per dozen.

PROCESS.

Net price, $17.÷85=$20.

Solution.—Asking price assumed, 100
    Less proposed discount, 15
            85)1700($20
               170

EXAMPLE II.—A wholesale and retail merchant wishes to mark his goods and wares at such a price as to enable him to offer them to his wholesale customers at a discount of 25%, and still gain 15% above first cost—what must be his asking or retail price for goods costing $12?

PROCESS.

Cost, 12.00+1.80=13.80÷75=$18.40 Retail price.

Solution.   *Proof.*

Cost,   12.00   18.40 Retail price.
Wholesale price 15% 1.80   4.60 25% discount.
Asking price,   13.80   13.80 Wholes'e price.
Retail price assumed, 100   1.80 15%prof. deduc.
Less proposed disc't,   25 $12.00 Cost.
                        75)138000(18.40
                           75

                           630
                           600

                           300
                           300

                             0

DISCOUNT FROM ASKING PRICE.

To ascertain the per cent. of discount from the asking price to produce the cost.

RULE.—*From the asking price subtract the cost, divide the difference by the asking price, and the quotient will be the per cent. of discount.*

EXAMPLE.—The asking price of a case of goods costing $120 was $150—what per cent. of discount can be offered to have the goods net cost? *Ans.* 20%.

Solution.   *Proof.*

$150 Asking price.   Asking price, $150
 120 Cost.   Discount 20% =   30
 ---   ----
 30÷150=20%.   Cost, $120

RAPID PROCESS OF MARKING GOODS.
A VALUABLE HINT TO MERCHANTS AND ALL RETAIL DEALERS IN FOREIGN AND DOMESTIC DRY GOODS.

RETAIL merchants, in buying goods by wholesale, buy a great many articles by the dozen, such as boots and shoes, hats and caps, and notions of various kinds. Now, the merchant, in buying, for instance, a dozen hats, knows exactly what one of those hats will retail for in the market where he deals; and, unless he is a good accountant, it will often take him some time to determine whether he can afford to purchase the dozen hats and make a living profit in selling them by the single hat; and in buying his goods by auction, as the merchant often does, he has not time to make the calculation before the goods are cried off. He therefore loses the chance of making good bargains by being afraid to bid at random, or if he bids, and the goods are cried off, he may have made a poor bargain, by bidding thus at a venture. It then becomes a useful and practical problem to determine *instantly* what per cent. he would gain if he retailed the hats at a certain price.

To tell what an article should retail for to make a profit of 20 per cent.,

RULE.—*Divide what the articles cost per dozen by 10, which is done by removing the decimal point one place to the left.*

For instance, if hats cost $17.50 per dozen, remove the decimal point one place to the left, making $1.75, what they should be sold for apiece to gain 20 per cent. on the cost. If they cost $31.00 per dozen, they should be sold at $3.10 apiece, etc. We take 20 per cent. as the basis for the following reasons, viz.: because we can determine instantly, by simply removing the decimal point, without changing a figure; and, if the goods would not bring at least 20 per cent. profit in the home market, the merchant could not afford to purchase and would look for goods at lower figures.

REASON.—The reason for the above rule is obvious: For if we divide the cost of a dozen by 12, we have the cost of a single article; then if we wish to make 20 per cent. on the cost, (cost being $\frac{1}{5}$ or $\frac{5}{5}$,), we add the 20 per cent., which is $\frac{1}{5}$, to the $\frac{5}{5}$, making $\frac{6}{5}$ or $\frac{12}{10}$; then as we multiply the cost, divided by 12, by the $\frac{12}{10}$ to find at what price one must be sold to gain 20 per cent., it is evident that the 12s will cancel, and leave the cost of a dozen to be divided by 10, which is done by removing the decimal point one place to the left.

1. If I buy 2 doz. caps at $7.50 per doz., what shall I retail them at to make 20%? Ans. 75 cts.

2. When a merchant retails a vest at $4.50 and makes 20%, what did he pay per doz.? Ans. $45

3. At what price should I retail a pair of boots that cost $85 per doz., to make 20%? Ans. $8.50.

RAPID PROCESS OF MARKING GOODS AT DIFFERENT PER CENTS.

Now, as removing the decimal point one place to the left, on the cost of a dozen articles, gives the selling price of a single one with 20 per cent. added to the cost, and, as the cost of any article is 100 per cent., it is obvious that the selling price would be 20 per cent. more, or 120 per cent.; hence, to find 50 per cent. profit, which would make the selling price 150 per cent., we would first find 120 per cent., then add 30 per cent., by increasing it one-fourth itself; to make 40 per cent., add 20 per cent., by increasing it one-sixth itself; for 35 per cent., increase it one-eighth itself, etc. Hence, to mark an article at any per cent profit, we have the following

GENERAL RULE

First find 20 *per cent. profit, by removing the decimal point one place to the left on the price the articles cost a dozen; then, as* 20 *per cent. profit is* 120 *per cent., add to or subtract from this amount the fractional part that the required per cent. added to* 100 *is more or less than* 120.

Merchants, in marking goods, generally take a per cent. that is an aliqot part of 100, as 25%, 33¼%, 50%, etc. The reason they do this is because it makes it much easier to add such a per cent. to the cost; for instance, a merchant could mark almost a dozen articles at 50 per cent. profit in the time it would take him to mark a single one at 49 per cent. For the benefit of the student, and for the convenience of business men in marking goods, we have arranged the following table:

TABLE

For Marking all Articles bought by the Dozen.

N. B. Most of these are used in business.

To make 20% remove the point one place to the left
" " 80% " " " and add $\frac{1}{2}$ itself.
" " 60% " " " " " $\frac{1}{3}$ "
" " 50% " " " " " $\frac{1}{4}$ "
" " 44% " " " " " $\frac{1}{5}$ "
" " 40% " " " " " $\frac{1}{5}$ "
" " 37½% " " " " " $\frac{1}{7}$ "
" " 35% " " " " " $\frac{1}{7}$ "
" " 33⅓% " " " " " $\frac{1}{8}$ "
" " 32% " " " " " $\frac{1}{16}$ "
" " 30% " " " " " $\frac{1}{13}$ "
" " 28% " " " " " $\frac{1}{18}$ "
" " 26% " " " " " $\frac{1}{20}$ "
" " 25% " " " " " $\frac{1}{25}$ "
" " 12½% " " " subtract $\frac{1}{16}$ "
" " 16⅔% " " " " $\frac{1}{25}$ "
" " 18¾% " " " " $\frac{1}{25}$ "

If I buy 1 doz. shirts for $28.00, what shall I retail them for to make 50%?   Ans. $3.50

EXPLANATION.—Remove the point one place to the left, and add on ¼ itself.

BASIS OF SUCCESS IN BUSINESS.

SHORT CREDITS SECURE LARGE PROFITS.

Be watchful of your credits, as they have much to do toward securing large profits in business.

The strong argument in favor of such a course can best be illustrated from the results shown in the following table, giving the accumulations from $100, invested in business for the term of ten years, and turned over at profits of 5, 8, and 10%.

TABLE.

Capital invested.					Amount, 5%.	Amount, 8%.	Amount, 10%
$100 turned over every			3 mos.		703.98	2172.46	4525.93
100	"	"	"	6 "	265.33	466.08	672.74
100	"	• "	"	12 "	162.87	215.88	259.36
100	"	"	"	2 years.	127.62	146.93	161.05

NOTE.—An important question to be considered in reducing prices is, can one's sales be sufficiently increased, so as to compensate for the reduction of profits? From the above table it will be observed that quick sales and

SMALL PROFITS

Are more desirable than large ones, when secured at the expense of long credits.

BEWARE OF EXPENSE.

As expenses diminish all profits it is essential that judicious economy be at all times exercised. Many a business becomes bankrupt from lack of foresight in this one particular.

LOOK TO YOUR CREDIT.

Always command a thorough knowledge of your business and your books.

A smaller business with cash capital produces larger profits than a large business conducted on credit.

Be not too anxious to extend your business or branch out.

Have a smaller house and larger capital.

Goods well bought and discounts saved secure early profits.

Avoid outside speculation, the chances are against success.

Do not overtrade: goods in stock are better than charged up in bad debts.

LEDGER ACCOUNTS

Ledger accounts arise from business transactions, and comprise a series of condensed statements under specific titles, showing one's relation with persons and property.

The left-hand side of each account shows debit entries, and the right-hand side credit entries.

CLASSIFICATION OF ACCOUNTS.

Accounts are divided or classified under two distinct heads, *i. e.*, *Real* and *Representative*.

Real Accounts are those exhibiting *Resources* or *Liabilities*.

Representative Accounts are those exhibiting *Gains* or *Losses*.

The several styles of accounts, such as are used in the most extensive business establishments, are herewith presented, showing in detail the office or exhibit of each. To the

LEDGER ACCOUNTS.

STUDENT OR ACCOUNTANT,

Who is desirous in *posting* himself regarding the principles governing accounts, and balancing or

CLOSING THE LEDGER,

The following work or illustrations will prove invaluable, and will amply repay for any time or labor expended, as it contains in a condensed form all that is of practical value appertaining to the subject.

Stock or Capital Account

Represents the proprietor, showing his capital or investment,

AT COMMENCING BUSINESS.

The Credit of Account shows the Capital or Resources invested.

The Debit of Account shows the Liabilities or Amounts assumed to be paid from the business.

The Difference shows the *Net Capital,* if the credit is in excess; and the *Net Insolvency,* if the debit is in excess.

DURING BUSINESS.

The Credit of Account will show any additional investments.

The Debit of Account will show any additional liabilities assumed, or withdrawals of capital.

UPON CLOSING THE LEDGER.

The Credit of Account will show the total investment of capital and the *Net Gain* arising from the business.

The Debit of Account will show the total liabilities assumed, withdrawals of capital, and *Net Loss* arising from the business.

The Difference will show the proprietor's *Present Worth* or *Insolvency*. If the credit side is in excess, a *Present Worth;* if the debit side is in excess, an *Insolvency*.

The account is *closed To or By Balance,* as shown by the excess.

Partner's Accounts

Are treated precisely like the stock or capital account. The *Net Interests of all the partners taken together* will exhibit the condition or standing of the firm.

Cash.

Classification, Real; Closed *By Balance.*

Debit of Account shows Total Cash Receipts.
Credit of Account shows Total Cash Payments.
The Difference shows a *Resource* consisting of the amount of *cash on hand.*

NOTE.—As more cash cannot be *paid out* than is *received,* the *credit* of the account *can never be in excess.*

Bills Receivable.

Classification, Real; Closed *By Balance.*

Debit of Account shows others' notes and acceptances received.

Credit of Account shows others' notes and acceptances paid, discounted or disposed of.

The Difference shows a *Resource* consisting of *others' paper on hand,* the value of which we are to receive.

NOTE.—As others' paper cannot be *disposed of until received,* the credit of the account can never be in excess.

Bills Payable.

Classification, Real; Closed *To Balance.*

Credit of Account shows our notes and acceptances issued.

Debit of Account shows our notes and acceptances that have been paid or redeemed.

The Difference shows a *Liability* consisting of *our outstanding* or unredeemed paper.

NOTE.—As we cannot *redeem or pay* more of our paper than is *issued*, the *debit* side of the account can never be in excess.

Personal Accounts.

Classification, Real; Closed *To or By Balance*.

Debit of Account shows our charges against them or indebtedness in our favor.

Credit of Account shows their charges against us, or our indebtedness in favor of others.

The Difference, if in favor of the *debit side*, shows a *Resource*, or the sum due us from the parties represented by the account.

If in favor of the *credit side*, a *Liability*, or amount we owe them.

NOTE.—When the balance is in our favor the account is termed a *Personal Account Receivable;* when against us a *Personal Account Payable.*

Consignments.

Accounts representing consignments of merchandise or property received from others, to be sold for their account and risk, are treated the same as *Personal Accounts.* The *debit side* of the account showing amount of expenses incurred, and the *credit side* the returns.

Merchandise.

Classification, Representative; Closed *To or By Loss and Gain.*

Debit of Account shows Cost of goods purchased.

Credit of Account shows Sales or Proceeds.

NOTE.—If the goods are not all sold or disposed of, it will be necessary to credit the account with the *inventory* or market value of the unsold quantity.

The Difference shows a *gain* or *loss* according to the *excess* of the sides. If in favor of the *credit*, a *Gain;* of the *debit*, a *Loss.*

NOTE.—This account may be made to comprise all properties purchased for traffic, such as groceries, drygoods, flour, produce, hardware, crockery, etc.; or, if desired, each kind may be represented under its own separate heading.

Shipments to _____

A special account under the above heading is frequently used to represent merchandise or property which we have consigned away to be sold for our account and risk. The account is treated similar to Merchandise. It is

Debited with the cost or outlay.

Credited with the returns.

NOTE.—At closing the Ledger, if account sales or fu... returns have not been received, it will be necessary to credit the account with the value represented, or inventory, same as in Merchandise.

The Difference shows a *Gain* or *Loss*. (See Merchandise account.)

Real Estate.

Classification, Representative; Closed *To or By Loss and Gain.*

Debit of Account shows Cost or Outlay.

Credit of Account shows Proceeds from Sales, or Income.

NOTE.—Add inventory, if any, same as in Merchandise.

The Difference shows *Gain* or *Loss* according to the *excess* of sides. (See Merchandise ac't.)

Store Fixtures.

Classification, Representative; Closed *By Loss and Gain.*

Debit of Account shows Cost or Outlay.

Credit of Account shows Returns, if any. If

value is estimated, add as Inventory in Merchandise.

The Difference will show a *Loss*.

Expense.

Classification, Representative; Closed *By Loss and Gain.*

Debit of Account shows Expenditures and outlay for conducting the business.

NOTE.—In case there should be returns from expense expenditures, the account would be credited.

The Difference shows a *Loss*.

Interest.

Classification, Representative; Closed *To or By Loss and Gain.*

Debit of Account shows sums paid for use of others' money.

Credit of Account shows sums received for use of our money.

The Difference, if in favor of the *credit* side, shows a *gain;* when in favor of the *debit* side, a *loss.*

Discount.

. Classification, Representative; Closed *To* or *By Loss and Gain.*

Debit of Account shows sums paid or allowed by us.

Credit of Account shows sums received or allowed us.

The Difference, if in favor of the *credit* side, shows a *gain;* when in favor of the *debit* side, a *loss.*

Insurance.

Classification, Representative; Closed *To* or *By Loss and Gain.*

Debit of Account shows amount paid for insurance.

Credit of Account shows amount received from others for insurance.

The Difference, if in favor of the *credit* side, shows a *gain;* when in favor of the *debit* side, a loss.

Commission.

Classification, Representative; Closed *To* or *By Loss and Gain.*

Debit of Account shows sums paid or allowed for services in buying or selling.

Credit of Account shows sums received or allowed for services in buying or selling

The Difference, if in favor of the *credit* side, shows a *gain;* when in favor of the *debit* side, a *loss.*

Loss and Gain

(or, *Profit and Loss*) exhibits Net Gains or Losses. Closed *To* or *By Stock* or *Capital,* wherein *single* proprietorship is represented. *In Partnership* business the account is closed *To* or *By Partners' Account,* showing each partner's respective share in the division of gains or losses.

Debit of Account shows the total *Losses* arising from the business.

Credit of Account shows the total *Gains* produced from the business.

The Difference shows a *Net Gain or Net Loss,* according to the excess of sides; a *Net Gain* if the difference is in favor of the *credit* side, and a *Net Loss* when in favor of the *debit* side.

NOTE.—In closing the account, the *Net Gain* is carried to the *credit* of *Stock* or *Partners'* account, showing an *increase* of *capital*. The *Net Loss* is carried to the *debit* of *Stock* or *Partners'* account, showing a *diminution* of *capital*.

Gains or *Losses* are divided between partners in accordance with the articles of copartnership or agreement made between them. The methods of adjustment will be found fully illustrated under the head of *Partnership*, page 139 of this work.

Private Account

Is a special account kept with the proprietor to show his transactions with the business as an individual.

The account is

DEBITED

With his withdrawals of money or the appropriation of any value belonging to the business to his private use, such as personal, household, or living expenses.

The account is closed *By Stock* or *Capital*, and total amount carried to the *debit* of the *Stock* or *Capital* account.

Partners' Private Accounts

Are special accounts kept with each partner showing their transactions with the firm as individuals. The accounts are

DEBITED

For precisely the same reasons as *explained* in the individual proprietor's *Private Account*.

They are closed *By Partners*, and each total *debit*, as shown, carried to the *debit* of the respective partner's *general* or *Capital Account*.

HOW TO CLOSE THE LEDGER.

Inventories.—Take an inventory of all unsold property and *credit* the *account*, showing the *cost* of same *when purchased*.

This entry is made with *red ink*, thus: *By Balance Inventory*, or *By Inventory*. After ruling, the amount is brought down to the *debit* of the *new account* in black ink.

Representative Accounts. — Those showing *Gains* or *Losses* are closed after crediting the inventory, if any, by writing on the smaller side, with *red ink*, *To Loss and Gain*, or *By Loss and*

CLOSING LEDGER ACCOUNTS.

Gain. The excess as shown is carried to the *debit* or *credit* of the *Loss and Gain* account, the entry being made in black ink.

Real Accounts.—Those showing *Resources* or *Liabilities* are closed by writing on the smaller side, in *red ink, To Balance* or *By Balance.*

After ruling, the balance of excess as shown is brought to the *debit* or *credit* of the new account.

Loss and Gain.—After closing all *Representative* accounts, and the excess of gains or losses having all been brought forward, write on the smaller side, in *red ink, To Stock or Partners'* or *By Stock or Partners'.* The difference will show the *Net Gain* or *Net Loss,* which carry to the *debit* or *credit* of Stock or Partners' account.

Stock or Partners' Account.—Write on the smaller side, in *red ink, To Balance* or *By Balance.* Rule up the account and bring down the balance as shown to the debit or credit of the new account in black ink.

Proof of Work.—Take a *Trial Balance* after closing, and if your work is correct, the **Ledger** will be in *equilibrium.*

ERRORS IN TRIAL BALANCES.

VALUABLE RULES FOR THEIR DETECTION.

From Bryant & Stratton's Business Arithmetic, with kind permission of the author, H. B. Bryant, President Bryant & Stratton Business College, Chicago, Illinois.

In keeping accounts by Double Entry, each item appears in at least two different accounts, on the Dr. side in one and on the Cr. side in the other; hence the sum of all the debit entries in all the accounts will equal the sum of all the credit entries in all the accounts, and the sum of the Dr. balances will equal the sum of the Cr. balances, if all the entries are properly made

ILLUSTRATION OF DOUBLE ENTRY.

MERCHANDISE.

Dr.					Cr.
Apr.	10	To Cash	$ 75	Apr. 1	By Cash	$3000
"	12	" "	380	May 10	" "	500
May	2	" "	1200			
"	15	" "	725			
		Balance	1120			
			$3500			$3500

CASH.

Dr.					Cr.
Apr.	1	To Mdse.	$3000	Apr. 10	By Mdse.	$ 75
May	10	" "	500	" 12	" "	380
				May 2	" "	1200
				" 15	" "	725
					Balance	1120
			$3500			$3500

A TRIAL BALANCE

Is a summary of the entire amounts entered on the Dr. and Cr. side of each account, or simply of the balances of all the accounts in detail, and if the sums of the debit and credit entries in the Trial Balance are not equal, then there is some error in the accounts or in making up the Trial Balance, which should be discovered and corrected.

The first rule of the book-keeper should be *to make no error;* but for such as are fallible the following suggestions may be of some practical utility.

1. If the error be found in one figure only, it is probably an error of adding or copying.

2. If it involve several figures, it may have arisen from the omission of an entire entry or from making the same entry twice.

3. If it be divisible by 2, without a remainder, it may have arisen from posting an item to the wrong side of the account, in which case the item would be half of the apparent error.

4. If the error be divisible by 9, without a remainder, it may have arisen from transposition, three cases of which may be easily detected by rules founded on the peculiar property of the number 9. These cases are—

1st. When two figures are made to exchange places with each other, the orders in notation remaining the same: *e. g.*, 372 made to read 327, or 732, or 273.

2d. When two or more figures are made to change their places in notation, their arrangement in respect to each other remaining the same: *e. g.*, $4275 made to read $42750, or $42.75, or $427.50.

3d. When *two* significant figures are made to change position both with respect to each other and also the orders of notation: *e. g.*, $14 made to read $0.41.

5. To detect the first and second cases of transposition *divide the amount of the error in the trial balance successively by* 9, 99, 999, 9999, *etc., as far as*

possible without a remainder, rejecting all ciphers at the right of the last significant figure in the error.

The quotients that contain but one digit figure will express the difference between the two digit figures transposed, which will be adjacent to each other if the divisor consist of but one 9, separated by one other figure if it consist of two 9's, by two other figures if it consist of three 9's, and so on.

Those quotients which contain two or more figures will express the *number* itself, which is transposed in notation simply, the arrangement of the significant figures remaining the same. In either case the *order* of the last *significant* figure in the error will be the lowest order of the figures transposed. The orders of the other figures can be easily determined by referring to the error and applying the principles of notation.

6. To detect the *third* case, divide the error in the balance by as many 9's as is possible so as to give only a single figure in the quotient, and then the remainder in the same way, rejecting all ciphers at the right of the last significant figure in both dividends, after which there should be no remainder.

The first quotient will be the figure filling both the highest and lowest order in the transposition; the second quotient will be the other figure.

NOTE —If the error of the trial balance be not divisible by 9 it cannot be the result of transposition alone. But whenever the error becomes so reduced as to be divisible by 9 without a remainder, a transposition being then possible, the above tests should be applied.

WE ADD THE FOLLOWING
IMPORTANT SUGGESTIONS.

First.—Examine the Cash and Bills Receivable accounts; the balance can never appear on the credit side, and should equal the amount of cash and notes on hand, as shown by the C. B. and B. B.

Second.—Refer to the Bills Payable account; the balance shown must appear in favor of the credit side and equal the amount of our outstanding paper.

Third.—If the cash book is kept as an original book of entry, see that the balance on hand from previous month has been deducted.

Fourth.—If the error appears only in the cent or dollar columns, it is not necessary to add the columns to the left.

Fifth.—If, after applying the above tests, the error still exists, it will be necessary to go over the entire work. Re-add carefully the debit and credit sides of all the accounts, as the error undoubtedly lies in the addition. If not found, examine each posting separately, check from the journal or book posted from to the ledger, and *vice versa*, as you proceed with your work.

NOTE.—In the use of these rules in practice, not only the balances of the ledger accounts as they appear on the balance-sheet should be examined, but also all the separate postings, as a transposition *there* will equally affect the final balance.

MEASUREMENT OF LUMBER.

The unit of board measure is a square foot 1 inch thick.

To measure inch boards.

RULE.—Multiply the length of the board in feet by its breadth in inches, and divide the product by 12; the quotient is the contents in square feet.

NOTE.—When the board is wider at one end than the other, add the width of the two ends together, and take half the sum for a mean width.

EXAMPLE.—How many square feet in a board 10 feet long, 13 inches wide at one end, and 9 inches wide at the other?

Process.—$(13 + 9) \div 2 = 11$ (mean width) then 10 length $\times 11 = 110 \div 12 = 9\frac{1}{6}$ feet. *Ans.*

Sawed lumber, as joists, plank, and scantlings, are now generally bought and sold by *board measure*. The dimensions of a *foot* of board measure are 1 foot long, 1 foot wide, and 1 inch thick.

To ascertain the contents (board measure) of boards, scantling, and plank.

RULE.—Multiply the width in *inches* by the thickness in *inches*, and that product by the length in *feet*, which last product divide by 12.

EXAMPLE.—How many feet of lumber in 14 planks 16 feet long, 18 inches wide and 4 inches thick?

Process.—16 feet $\times$ 18 inches $\times$ 4 inches $= 1152$, then $1152 \div 12 = 96$ feet $=$ contents of one plank. $96 \times 14 = 1344$ feet. *Ans.*

To find how many feet of lumber can be sawed from a log. (Gauge of saw $\frac{1}{4}$ inch.)

RULE.—From the diameter of the log in *inches*, subtract 4 (for slabs), one-fourth of this remainder squared and multiplied by the length in *feet* will give the correct amount of lumber that can be made from any log whatever.

EXAMPLE.—How many feet of lumber can be made from a log which is 36 inches in diameter, and 10 feet long?

Process.—From 36 (diameter) subtract 4 (for slabs) $= 32$, then divide the 32 by 4, making 8, which squared $= 64$, then multiply the 64 by 10 (length) $= 640$ feet. *Ans.*

MEASUREMENT OF LUMBER.

To find how many feet of lumber there are left in a log after it is made perfectly square.

RULE.—Multiply the diameter in *inches* at the small end by one-half the number of inches, and this product by the length of the log in *feet*, which last product divide by 12.

EXAMPLE.—If the diameter of a round stick of timber be 22 inches, and its length 20 feet, how much lumber will it contain when hewn square?

Half diameter $= 11$, and $\dfrac{22 \times 11 \times 20}{12} = 403\frac{1}{3}$ ft., the lumber when hewn square.

To find how many feet of square edged boards, of a given thickness, can be sawed from a log of a given diameter.

RULE.—Find the quantity of lumber in the log, when made square, by the last Rule; then divide by the thickness of the board, including the saw calf, the quotient is the number of feet of boards.

EXAMPLE.—How many feet of square edged boards, $1\frac{1}{4}$ inch thick, including the saw calf, can be sawn from a log 20 feet long, and 24 inches diameter?

$$\dfrac{24 \times 12 \times 20}{12} = 480 \text{ ft.}, \text{ the lumber when hewn sq.}$$

Then 480 divided by $1\frac{1}{4} = 384$ feet. *Ans.*

MEASUREMENT OF WOOD.

Wood is measured by the cord, which contains 128 cubic feet.

Wood is bought and sold by the cord and fractions of a cord.

Pine and spruce spars from 10 to 4 inches in diameter inclusive, are measured by taking the diameter, clear of bark, at one-third of their length from the large end.

Spars are usually purchased by the inch diameter; all under 4 inches are considered *poles*.

Spruce spars of 7 inches and less, should have 5 feet in length for every inch in diameter.

NOTE.—A pile of wood that is 8 feet long, 4 feet high, and 4 feet wide, contains 128 cubic feet, or a cord, and every cord contains 8 *cord-feet;* and as 8 is $\frac{1}{16}$ of 128, every *cord-foot* contains 16 cubic feet; therefore, dividing the cubic feet in a pile of wood by 16, the quotient is the cord-feet; and if cord-feet be divided by 8, the quotient is cords.

NOTE.—If we wish to find the circumference of a tree which will hew any given number of inches square, we divide the given side of the square by .225, and the quotient is the circumference required.

What must be the circumference of a tree that will make a beam 10 inches square?

NOTE.—When wood is "corded" in a pile 4 feet wide, by multiplying its length by its hight, and dividing the product by 4, the quotient is the cord-feet; and if a load of wood be 8 feet long, and its hight be multiplied by its width, and the product divided by 2, the quotient is the cord-feet.

How many cords of wood in a pile 4 feet wide, 10 feet 6 inches long, and 5 feet 3 inches high?

NOTE.—Small fractions rejected.

To find how large a cube may be cut from an given sphere, or be inscribed in it.

RULE.—*Square the diameter of the sphere, divide that product by 3, and extract the square root of the quotient for the answer.*

I have a piece of timber, 30 inches in diameter; how large a square stick can be hewn from it?

RULE.—*Multiply the diameter by .7071, and the product is the side of a square inscribed.*

I have a circular field, 360 rods in circumference, what must be the side of a square field that shall contain the same quantity?

RULE.—*Multiply the circumference by .282, and the product is the side of an equal square.*

I have a round field, 50 rods in diameter; what is the side of a square field that shall contain the same area?   Ans. 44.31135+rods.

RULE.—*Multiply the diameter by .886, and the product is the side of an equal square.*

There is a certain piece of round timber, 30 inches in diameter; required the side of an equilateral triangular beam that may be hewn from it.

RULE.—*Multiply the diameter by .866, and the product is the side of an inscribed equilateral triangle.*

To find the area of a globe or sphere.

DEFINITION.—A sphere or globe is a round solid body, in the middle or center of which is an imaginary point, from which every part of the surface is equally distant. An apple, or a ball used by children in some of their pastimes, may be called a sphere or globe.

ROUND TIMBER.

Round timber, when squared, is estimated to lose *one-fifth;* hence (50 cubic feet, or) a ton of round timber is said to contain only 40 cubic feet.

Round, sawed, and hewn timber is bought and sold by the cubic foot.

To measure round timber.

RULE.*—Take the girth in feet, at both the large and small ends, add them, and divide their sum by two for the mean girth; then multiply the length in feet by the square of one-fourth of the mean girth, and the quotient will be the contents in cubic feet, *according to the common practice.*

*This rule gives about *four-fifths* of the true contents, *one-fifth* being allowed to the buyer for waste in hewing.

EXAMPLE.—What are the cubic contents of a round log 20 feet long, 9 feet girth at the large end, and 7 feet at the small end?

SOLUTION.—$9 + 7 = 16 \div 2 = 8$ mean girth.

Then 20 length × 4 feet (the square of ¼ mean girth) = 80 cubic feet. *Ans.*

Note.—If the girth be taken in inches, and the length in feet, divide the last product by 144.

EXAMPLE.—What are the cubic contents of a round log 12 feet long, 50 inches girth at the large end, 38 inches at the small end?

WORK.—$50 + 38 = 88 \div 2 = 44$ mean girth.

Then 12 length × 121 inches (the square of ¼ mean girth) = $1452 \div 144 = 10\frac{1}{12}$ cubic feet.

To measure round timber as the frustum of a cone: that is, to measure all the timber in the log.

RULE.—Multiply the square of the circumference at the middle of the log in feet by 8 times the length, and the product divided by 100 will be the contents. *Extremely near the truth.*

NOTE.—The above rule makes 1 foot more timber in every 190 cubic feet a log contains if ciphered out by the long and tedious rules of Geometry. It is therefore sufficiently correct for all practical purposes, and *this* rule being so short and simple in comparison with *all others*, every lumberman, shipbuilder, carpenter, inspector or surveyor of timber, should post it up for reference and use.

A TABLE FOR MEASURING TIMBER.

Quarter Girt.	Area.	Quarter Girt.	Area.	Quarter Girt.	Area.
Inches	Feet.	Inches.	Feet.	Inches.	Feet.
6	.250	12	1.000	18	2.250
6¼	.272	12¼	1.042	18½	2.376
6½	.294	12½	1.085	19	2.506
6¾	.317	12¾	1.129	19½	2.640
7	.340	13	1.174	20	2.777
7¼	.364	13¼	1.219	20½	2.917
7½	.390	13½	1.265	21	3.062
7¾	.417	13¾	1.313	21½	3.209
8	.444	14	1.361	22	3.362
8¼	.472	14¼	1.410	22½	3.516
8½	.501	14½	1.460	23	3.673
8¾	.531	14¾	1.511	23½	3.835
9	.562	15	1.562	24	4.000
9¼	.594	15¼	1.615	24½	4.168
9½	.626	15½	1.668	25	4.340
9¾	.659	15¾	1.722	25½	4.516
10	.694	16	1.777	26	4.694
10¼	.730	16¼	1.833	26½	4.876
10½	.766	16½	1.890	27	5.062
10¾	.803	16¾	1.948	27½	5.252
11	.840	17	2.006	28	5.444
11¼	.878	17¼	2.066	28½	5.640
11½	.918	17½	2.126	29	5.840
11¾	.959	17¾	2.187	29½	6.044
				30	6.250

To measure round timber by the table.

Multiply the area corresponding to the **quarter girt** in *inches* by the length of the log in *feet*.

Note.—If the quarter-girt exceed the table, take half of it, and four times the contents thus formed will be the answer.

EXAMPLE 1.

If a piece of round timber be 18 feet long, and the quarter girt 24 inches, how many feet of timber are contained therein?

```
    24 quarter girt.
    24
    ──
    96
    48
    ───
   576 square.         By the Table.
    18
    ────           Against 24 stands  4.00
   4608                     Length,    18
    576                              ─────
─────────           Product,         72.00
144)10368(72 feet
    1008                 Ans. 72 feet.
    ────
     288
     288
```

This table gives the customary, but only about *four-fifths* of the true contents, *one-fifth* being allowed the buyer for waste in hewing or sawing to make the timber square.

The following rule gives the true contents:—

Multiply square of girth by .08 times length In the above example the whole girth is 8 feet. squared is 64 × (.08 × 18 length) = 92.16 feet.

1. *Of Flooring.*

Joists are measured by multiplying their breadth by their depth, and that product by their length. They receive various names, according to the position in which they are laid to form a floor, such as trimming joists, common joists, girders, binding joists, bridging joists and ceiling joists.

Girders and joists of floors, designed to bear great weights, should be let into the walls at each end about two-thirds of the wall's thickness.

In boarded flooring, the dimensions must be taken to the extreme parts, and the number of squares of 100 feet must be calculated from these dimensions. Deductions must be made for staircases, chimneys, etc.

Example 1. If a floor be 57 feet 3 inches long, and 28 feet 6 inches broad, how many squares of flooring are there in that room?

By Decimals.	*By Duodecimals*
57.25	F. I.
28.5	57 : 3
	28 : 6
28625	
45800	456
11450 .	114
	28 : 7 : 6
100)1631.625 feet.	7 : 0 : 0
Squares 16.31625	16:31 : 7 : 6

Ans. 16 squares and 31 feet.

SQUARE TIMBER.

To measure square timber.

RULE.—Multiply the breadth in feet by the depth in feet, and that by the length in feet, and the quotient will be the contents in cubic feet.

EXAMPLE.—How many cubic feet in a square log 12 feet long by 2 feet broad and 1½ feet deep?

EXPLANATION.—2 feet breadth × 1½ feet depth × 12 feet length = 36 cubic feet. *Ans.*

Note.—If the breadth and depth be taken in inches, divide the last product by 144.

EXAMPLE.—How many cubic feet in a square log 24 feet long, 30 inches broad, and 20 inches deep?

SOLUTION.—30 inches breadth × 20 inches depth × 24 feet length = 14400 ÷ 144 = 100 cubic feet

PROBLEM.

To find the solid contents of squared or four-sided Timber.

By the Carpenters' Rule.

As 12 on D: length on C: Quarter girt on D: solidity on C.

RULE I.—*Multiply the breadth in the middle by the depth in the middle, and that product by the length for the solidity.*

NOTE.—If the tree taper regularly from one end to the other, half the sum of the breadths of the two ends will be the breadth in the middle, and half the sum of the depths of the two ends will be the depth in the middle.

RULE II.—*Multiply the sum of the breadths of the two ends by the sum of the depths, to which add the product of the breadth and depth of each end; one-sixth of this sum multiplied by the length will give the correct solidity of any piece of squared timber tapering regularly.*

PROBLEM.

To find how much in length will make a solid foot or any other assigned quantity of squared timber of equal dimensions from end to end.

RULE.—*Divide 1728, the solid inches in a foot, or the solidity to be cut off, by the area of the end in inches, and the quotient will be the length in inches.*

NOTE.—To answer the purpose of the above rule, some carpenters' rules have a little table upon them, in the following form, called a *table of timber measure.*

0	0	0	0	9	0	11	3	9	inches.
144	36	16	9	5	4	2	2	1	feet.
1	2	3	4	5	6	7	8	9	side of the square.

This table shows, that if the side of the square be 1 inch, the length must be 144 feet; if 2 inches be the side of the square, the length must be 36 feet, to make a solid foot.

CAPACITY OF CISTERNS OR WELLS.

Tabular view of the number of gallons contained in the clear between the brickwork for each ten inches of depth:

Diameter.	Gallons.	Diameter.	Gallons.
2 feet equal..............	19	8 feet equal..............	313
2½ " "	30	8½ " "	353
3 " "	44	9 " "	396
3½ " "	60	9½ " "	461
4 " "	78	10 " "	489
4½ " "	97	11 " "	592
5 " "	122	12 " "	705
5½ " "	148	13 " "	827
6 " "	176	14 " "	959
6½ " "	207	15 " "	1101
7 " "	240	20 " "	1958
7½ " "	275	25 " "	2150

CISTERNS AND RESERVOIRS.
How to measure their contents.

1st. Find the solid contents of the cistern in cubic inches.

2d. Divide the contents so found by 14553, and the quotient will be the number of hogsheads.

If the height of the cistern be given, how do you find the diameter, so that the cistern shall contain a given number of hogsheads?

1st. Reduce the height of the cistern to inches, and the contents to cubic inches.

2d. Multiply the height by the decimal .7854.

3d. Divide the contents by the last result, and extract the square root of the quotient, which will be the diameter of the cistern in inches.

NOTE.—In estimating the capacity of cisterns, reservoirs, etc., the following table is used:

31½ gallons............................1 barrel.
63   "   ....1 hogshead.

The barrels used in commerce vary from 30 to 45 gallons, and the hogshead from 40 to 60 gallons.

CISTERNS AND RESERVOIRS.

HOW TO MEASURE THEIR CONTENTS.

Cisterns and Reservoirs are constructed for the purpose of holding large quantities of water or fluids, and are in the form of a tub cylinder, or solid square. They are generally built in the ground, and mortised on all sides, except the opening, with brick or stone, and are chiefly permanent in their construction.

TO MEASURE ROUND OR CYLINDRICAL CISTERNS.

RULE I.—*Multiply the square of the diameter in feet by the depth in feet, which will give the number of cylindrical feet in the cistern.*

RULE II.—*Multiply the cylindrical feet by*

$$\frac{373}{4000} \text{ for Hogsheads.}$$

$$\frac{373}{2000} \text{ for Barrels, or divide by 5,}$$

$$\frac{47}{8} \text{ for Gallons.}$$

The result will be the contents either in hogsheads, barrels, or gallons, depending upon the fractions used.

EXAMPLE I.—What is the contents in hogsheads of a cistern 20 feet in diameter and 10 feet deep?

SOLUTION.—20 ft. × 20 ft. = 400 ft. square of diameter.
400 " × 10 " = 4000 cylindrical ft.

$$4000 \text{ cylindrical ft.} \times \frac{373}{4000} = 373 \text{ hhds.}—Ans.$$

EXAMPLE II.—How many gallons in a cistern 10 feet in diameter and 16 feet deep?

TO MEASURE CISTERNS.

SOLUTION.—$10 \times 10 = 100 \times 16 = 1600$ cylindrical ft.

$$1600 \times \frac{47}{8} = 9400 \text{ gallons.}—Ans.$$

Or,
$$1600 \times \frac{373}{2000} = 298\tfrac{2}{5} \text{ barrels.}—Ans.$$

Or,
$$1600 \times \frac{373}{4000} = 149\tfrac{1}{5} \text{ hogsheads.}—Ans.$$

TO MEASURE SQUARE CISTERNS.

RULE I.—*Multiply the width in feet by the length in feet, and that product by the depth in feet; this last product will be the number of cubic feet in the cistern.*

RULE II.—*Multiply the cubic feet thus obtained by*

$$\frac{19}{160} \text{ for Hogsheads.}$$

$$\frac{19}{80} \text{ for Barrels.}$$

$$7\tfrac{48}{100} \text{ for Gallons.}$$

This will give the contents either in hogsheads, barrels, or gallons, as desired.

EXAMPLE.—How much water will a cistern contain which is 6 feet wide, 8 feet long, and 10 feet deep?

SOLUTION.—$6 \times 8 = 48 \times 10 = 480$ cubic ft. in cistern.

$$480 \times \frac{19}{160} = 57 \text{ hogsheads.}—Ans.$$

Or,
$$480 \times \frac{19}{80} = 114 \text{ barrels.}—Ans.$$

Or,
$$480 \times 7\tfrac{48}{100} \text{ or } 480 \times 7.48 = 3590\tfrac{40}{100} \text{ gall.}—Ans.$$

CASK-GAUGING.

Gauging is the art of measuring the capacity of casks and vessels of any form. In commerce, most of the gauging is done by the use of *the diagonal rod*, which gives only approximate results, but sufficiently accurate for ordinary purposes.

Ullage is the difference between the actual contents of a vessel and its capacity, or that part which is empty.

To measure small cylindrical vessels.

RULE.—Multiply the square of the diameter, in inches, by 34, and that by the height, in inches, and point off four figures; the result will be the capacity, in wine gallons and decimals of a gallon

For beer gallons multiply by 28 instead of 34.

EXAMPLE.—A can measures 15 inches in diameter, and is 2 feet 2 inches in height. How many gallons will it contain? $15 \times 15 = 225 \times 26$ height $= 5850$; $5850 \times 34 = 19.8900$. Ans. $19\frac{89}{100}$ galls.

Casks are usually regarded as the two equal frustums of a cone, and are very accurately gauged by three dimensions as follows:—

To measure a cask by three dimensions.

1st. Add the bung and head diameters in inches, and divide by 2 for the *mean* diameter.

2d. Multiply the square of the mean diameter by the length of the cask in inches.

3d. Multiply the last product by .0034 for wine gallons, .0028 for beer gallons.

EXAMPLE.—How many wine gallons in a cask, the bung diameter of which is 22 inches, the head diameter 20 inches, and the length 32 inches?

WORK.—$22 + 20 = 42 \div 2 = 21$ (mean diameter): then $21 \times 21 = 441$ (square of mean diameter), $\times 32$ length $= 14112 \times .0034 = 47.9808$. *Ans*

Note.—If the cask is not full, stand it on the end, and multiply by the *height of the liquid*, instead of the length of the cask, for actual contents.

When the cask is much bilged or rounded from the bung to the head, a more accurate way is to gauge by four dimensions, as follows:—

To measure a cask by four dimensions.

1st. Add the bung and head diameters in inches and the diameter in inches between bung and head

2d. Divide their sum by 3 for the *mean* diameter

3d. Multiply the square of the mean diameter by the length of the cask in inches.

4th. Multiply the last product by .0034 for wine gallons, .0028 for beer gallons.

EXAMPLE.—What are the contents in gallons of a cask, the bung diameter of which is 24 inches, the middle diameter 20 inches, the head diameter 16 inches, and its length 40 inches?

WORK.—$24 + 20 + 16 = 60 \div 3 = 20$ (mean diameter), then $20 \times 20 = 400$ (square of mean diameter) $\times 40$ length $= 16000 \times .0034 = 54.4$ gallons

1. The ale gallon contains 282 cubic inches.
2. The wine gallon contains 231 cubic inches.
3. The bushel contains 2150.4 cubic inches.
4. A cubic foot of pure water weighs 1000 ounces $= 62\frac{1}{2}$ pounds avoirdupois.
5. To find what weight of water may be put into a given vessel.

Multiply the cubic feet by 1000 for the ounces or by $62\frac{1}{2}$ for the pounds avoirdupois.

6. What weight of water can be put into a cistern $7\frac{1}{4}$ feet square? *Ans.* 26,367 lbs. 3 oz

MEASURING GRAIN.

By the United States standard, 2150 cubic inches make a bushel. Now, as a cubic foot contains 1728 cubic inches, a bushel is to a cubic foot as 2150 to 1728; or, for practical purposes, as 4 to 5. Therefore, to convert cubic feet to bushels, it is necessary only to multiply by $\frac{4}{5}$ or .8.

To measure the bushels of grain in a granary

RULE.—Multiply the length in *feet* by the breadth in feet, and that again by the depth in feet, and that again by $\frac{4}{5}$. The last product will be the number of bushels the granary contains.

EXAMPLE.—How many bushels in a bin 10 feet long, 4 feet wide, and 4 feet deep.

WORK.—10 feet length × 4 feet breadth × 4 feet depth = 160 cubic feet; then 160 × $\frac{4}{5}$ = 128. *Ans*

SIZE OF BINS

To contain a given number of bushels.

Having any number of bushels, how then will you find the corresponding number of cubic feet? Increase the number of bushels one-fourth itself, and the result will be the number of cubic feet.

How will you find the number of bushels which a bin of a given size will hold?

Find the content of the bin in cubic feet; then diminish the content by one-fifth, and the result will be the content in bushels.

How will you find the dimensions of a bin which shall contain a given number of bushels?

Increase the number of bushels one-fourth itself, and the result will show the number of cubic feet which the bin will contain. Then, when two dimensions of the bin are known, divide the last result by their product, and the quotient will be the other dimension.

If you wish the contents of a pile of ears of corn, or roots, in heaped bushels, ascertain the cubic feet, and multiply by $\frac{63}{100}$.

WEIGHTS AND MEASURES

Recognized by the Laws of the United States.

In some States dried apples and peaches are sold by the heaping bushel as are other of farm products.

A bushel of corn in the ear is three heaped half-bushels, or four even-full.

TABLE OF AVOIRDUPOIS POUNDS IN A BUSHEL,

As prescribed by statute in the several States named.

COMMODITIES	Cal.	Conn.	Ill.	Ind.	Iowa	Ky.	La.	Maine	Mass.	Mich.	Minn.	Mo.	N.J.	N.Y.	Ohio	Oregon	Penn.	R.I.	Vt.	Wis.
Barley	50	.	48	48	48	48	32	.	46	48	48	48	48	48	48	46	47	.	46	48
Beans	.	.	60	60	60	60	.	.	.	.	.	60	.	62	.	.	.	.	.	.
Blue Grass Seed	.	.	14	14	14	14	.	.	.	.	.	14	.	.	.	.	.	.	.	.
Buckwheat	40	45	46	50	52	52	.	.	46	42	42	52	50	48	.	42	48	.	46	42
Castor Beans	.	.	40	46	46	.	.	.	.	.	42	46	.	.	.	.	.	.	.	.
Clover Seed	.	.	60	60	60	60	.	.	.	60	60	60	64	60	60	60	.	.	.	60
Dried Apples	.	.	24	25	24	.	.	.	.	28	28	24	.	.	.	28	.	.	.	28
Dried Peaches	.	.	33	33	33	.	.	.	.	28	28	33	.	.	.	.	.	.	.	28
Flax Seed	.	.	56	56	56	56	.	.	.	.	.	56	55	.	56	56	.	.	.	56
Hemp Seed	.	.	44	44	44	44	.	.	.	.	.	14	55	55	.	.	.	.	.	.
Indian Corn	.	56	52	56	56	56	56	.	56	56	56	52	56	58	56	56	56	50	56	56
Indian Corn in ear	52	.	70	68	.	50	.	50	50	50	50	.	.	.	.	.	.	.	.	.
Indian Corn Meal	.	28	48	50	.	35⅓	.	30	30	.	.	.	.	.	.	.	.	50	.	.
Oats	32	.	32	32	35	57	32	.	52	32	32	35	30	32	32	34	32	.	32	32
Onions	.	.	57	48	57	.	.	60	60	.	.	57	.	.	.	.	.	50	.	.
Potatoes	.	60	60	60	60	60	.	50	56	60	60	60	60	60	60	60	60	60	60	60
Rye	54	.	60	56	56	56	.	.	50	56	56	56	56	56	56	56	56	50	56	60
Rye Meal	.	56	51	.	.	.	.	.	.	.	.	.	.	56	.	.	.	.	.	.
Salt	.	.	50	50	50	50	32	.	.	.	.	50	.	.	.	.	.	.	.	.
Timothy Seed	.	.	45	45	45	45	.	.	.	.	.	45	.	44	.	.	.	.	.	.
Wheat	60	.	60	60	60	60	60	.	60	60	60	60	60	60	60	60	60	.	60	46
Wheat Bran	.	.	20	.	20	20	.	.	.	.	.	20	.	.	.	.	.	.	.	60

In Pennsylvania 80 lbs. coarse, 70 lbs. ground, or 62 lbs. fine salt make 1 bushel; and in Illinois, 50 lbs. common, or 55 lbs. fine salt make 1 bushel. In Tennessee 100 ears of corn are a bushel. A heaping bushel contains 2815 cubic inches.

In Maine 64 lbs. of ruta baga turnips or beets make 1 bushel.

A cask of lime is 240 lbs. Lime in slacking absorbs 2½ times its volume, and 2¼ times its weight in water.

RAILROAD FREIGHT.
TABLE OF GROSS WEIGHTS.

The Articles named are Billed at actual weights, if possible, but usually at the weights in the Table below when it is not convenient to weigh them.

Ale and Beer.....320 lb. per bbl.	Highwines........ 350 lb. per bbl.
" " ....170 " ½ "	Hungarian Grass
" " ....100 " ¼ "	Seed.......... 45 " bu.
Apples, dried.... 24 " bu.	Lime............ 200 " bbl.
" green.... 56 " "	Malt............ 38 " bu.
" " ....150 " bbl.	Millet.......... 45 " "
Barley.......... 48 " bu.	Nails........... 108 " keg.
Beans, white..... 60 " "	Oats............ 32 " bu.
" castor..... 46 " "	Oil............. 400 " bbl.
Beef...........320 " bbl.	Onions.......... 57 " bu.
Bran............ 20 " bu.	Peaches, dried... 33 " "
Brooms.......... 40 " doz.	Pork............ 320 " bbl.
Buckwheat...... 52 " bu.	Potatoes, common. 150 " "
Cider..........350 " bbl.	" " . 60 " bu.
Charcoal........ 22 " bu.	" sweet... 55 " "
Clover Seed..... 60 " "	Rye............. 56 " "
Corn............ 56 " "	Salt, fine....... 56 " "
" in ear...... 70 " "	" "........ 300 " bbl.
" Meal....... 48 " "	" coarse...... 350 " "
" " ....220 " bbl.	" in sacks.... 200 " sack.
Eggs...........200 " "	Timothy Seed.... 45 " bu.
Fish...........300 " "	Turnips......... 56 " "
Flax Seed....... 56 " bu.	Vinegar......... 350 " bbl.
Flour..........200 " bbl.	Wheat........... 60 " bu.
Hemp Seed..... 44 " bu.	Whiskey......... 350 " bbl.

One *ton* weight is 2000 lbs.

ESTIMATED WEIGHTS OF LUMBER AND OTHER ARTICLES.

NOTE.—From 18,000 to 20,000 lb. is considered a car-load in most places, each car *itself* also weighing about 20,000 lb.

TO MEASURE CORN ON THE COB IN CRIBS.

Corn is generally put up in cribs made of rails; but the rule will apply to a crib of any size or kind, whether equilateral, or flared at the sides.

When the crib is equilateral

RULE.—Multiply the length in feet by the breadth in feet, and that again by the height in feet, which last product multiply by .63 (the fractional part of a heaped bushel in a cubic foot), and the result will be the heaped bushels of ears. For the number of bushels of shelled corn multiply by 42 (two-thirds of 63), instead of .63.

EXAMPLE.—Required the number of bushels of shelled corn contained in a crib of ears, 15 feet long, by 5 feet wide, and 10 feet high?

15 length × 5 width, × 10 height = 750 cubic feet. Then 750 × .63 = 472.50 heaped bushels of ears. Also 750 × .42 = 315 bushels of shelled corn.

In measuring the height, of course, the height of the corn is intended. And there will be found to be a difference in measuring corn in this mode, between fall and spring, because it shrinks very much in the winter and spring, and settles down.

When the crib is flared at the sides.

RULE.—Multiply half the sum of the top and bottom widths in *feet* by the perpendicular height in *feet*, and that again by the length in feet, which last product multiply by .63 for heaped bushels of ears, and by .42 for the number of bushels of shelled corn.

Note.—The above rule assumes that three heaping half bushels of ears make one struck bushel of shelled corn. This proportion has been adopted upon the authority of the major part of our best agricultural journals. Nevertheless, some journals claim that two heaping bushels of ears to one of shelled corn is a more correct proportion, and it is the custom in many parts of the country to buy and sell at that rate. Of course much will depend upon the kind of corn, the shape of the ear, the size of the cob, &c. Some samples are to be found, three heaping half bushels of which will even overrun one bushel shelled; while others again are to be found, two bushels of which will fall short of one bushel shelled. Every farmer must judge for himself, from the sample on hand, whether to allow one and a half or two bushels of ears to one of shelled corn. In either case, it is only an approximate measurement, but sufficient for all ordinary purposes of estimation. The only true way of measuring all such products is by weight.

MEASUREMENT OF HAY.

The only correct mode of measuring hay is to weigh it. This, on account of its bulk and character, is very difficult, unless it is baled or otherwise compacted. This difficulty has led farmers to estimate the weight by the bulk or cubic contents, a mode which is only approximately correct. Some kinds of hay are light, while others are heavy, their equal bulks varying in weight. But for all ordinary farming purposes of estimating the amount of hay in meadows, mows, and stacks, the following rules will be found sufficient:—

As nearly as can be ascertained, 25 cubic yards of average meadow hay, in windrows, make a ton

MEASUREMENT OF HAY.

When loaded on wagons, or stored in barns, 20 cubic yards make a ton.

When well settled in mows, or stacks, 15 cubic yards make a ton

Note.—These estimates are for medium-sized mows or stacks; if the hay is piled to a great height, as it often is where horse hay-forks are used, the row will be much heavier per cubic yard.

When hay is *baled*, or closely packed for shipping, 10 cubic yards will weigh a ton.

To find the number of tons in long square stacks.

RULE.—Multiply the length in yards by the width in yards, and that by *half* the altitude in yards, and divide the product by 15.

EXAMPLE.—How many tons of hay in a square stack 10 yards long, 5 wide, and 9 high?

SOLUTION.—$10 \times 5 \times 4\frac{1}{2} = 225 \div 15 = 15$ tons. *Ans.*

To find the number of tons in circular stacks.

RULE.—Multiply the square of the circumference in yards by 4 times the altitude in yards, and divide by 100; the quotient will be the number of cubic yards in the stack; then divide by 15 for the number of tons.

EXAMPLE.—How many tons of hay in a circular stack, whose circumference at the base is 25 yards, and height 9 yards?

SOLUTION.—25 × 25 = 625, the square of the circumference; then 625 × 36 (four times the length), = 225000 ÷ 100 = 225 (the number of cubic yards), then 225 ÷ 15 = 15, the number of tons.

An easy mode of ascertaining the value of a given number of lbs. of hay, at a given price per ton of 2000 lbs.

RULE.—Multiply the number of pounds of hay (coal, or anything else which is bought and sold by the ton) by one-half the price per ton, pointing off three figures from the right hand; the remaining figures will be the price of the hay (or any article by the ton).

EXAMPLE.—What will 658 lbs. of hay cost, @ $7 50 per ton?

SOLUTION.—$7 50 divided by 2 equals $3 75, by which multiply the number of pounds, thus: 658 × $3 75 = 246.750, or $2 46. *Ans*

Note.—The principle in this rule is the same as in interest—dividing the price by two gives us the price of half a ton, or 1000 lbs.; and pointing off three figures to the right is dividing by 1000.

A *truss* of hay, new, is 60 lbs.; old, 56 lbs; straw, 40 lbs.

A *load* of hay is 36 trusses.

A *bale* of hay is 300 lbs.

RULES FOR DETERMINING THE WEIGHT OF LIVE CATTLE.

For cattle of a girth of from 5 to 7 feet, allow 23 lbs. to the superficial foot.

For cattle of a girth of from 7 to 9 feet, allow 31 lbs. to the superficial foot.

For small cattle and calves of a girth of from 3 to 5 feet, allow 16 lbs. to the superficial foot.

For pigs, sheep, and all cattle measuring less than 3 feet girth, allow 11 lbs. to the superficial foot.

Measure in inches the girth round the breast, just behind the shoulder-blade, and the length of the back from the tail to the forepart of the shoulder-blade. Multiply the girth by the length, and divide by 144 for the superficial feet, and then multiply the superficial feet by the number of lbs. allowed for cattle of different girths, and the product will be the number of lbs. of beef, veal, or pork in the four quarters of the animal. To find the number of stone, divide the number of lbs. by 14.

EXAMPLE.—What is the estimated weight of beef in a steer, whose girth is 6 feet 4 inches, and length 5 feet 3 inches?

SOLUTION.—76 inches girth, × 63 inches length, = 4788 ÷ 144 = $33\frac{1}{4}$ square feet, × 23 = $764\frac{3}{4}$ lbs., or $54\frac{5}{8}$ stone. *Ans.*

Note.—When the animal is but half fattened, a deduction of one lb. in every 20 must be made; and if very fat, one lb. for every 20 must be added.

Where great numbers of cattle are annually bought and sold under circumstances that forbid ascertaining their weight with positive accuracy, the estimated weight may be thus taken with approximate exactness—at least with as much accuracy as is necessary in the aggregate valuation of stock. No rules or tables can, however, be at all times implicitly relied on, as there are many circumstances connected with the build of the animal, the mode of fattening, its condition, breed, &c., that will influence the measurement, and consequently the weight. A person skilled in estimating the weight of stock soon learns, however, to make allowance for all these circumstances.

BRICK BUILDING.

A perch of stone is 24.75 cubic feet; when built in the wall, 22 cubic feet make 1 perch, 2¾ cubic feet being allowed for the mortar and filling.

Three pecks of lime and four bushels of sand to a perch of wall.

To find the number of perches of stone in walls.

RULE.—Multiply the length in feet by the height in feet, and that by the thickness in feet, and divide the product by 22.

EXAMPLE.—How many perches of stone contained in a wall 40 feet long, 20 feet high, and 18 inches thick?

SOLUTION.—40 feet length × 20 feet height × 1½ feet thick = 1200 ÷ 22 = 54.54 perches   *Ans*

Note.—To find the perches of *masonry*, divide the cubic feet by 24.75, instead of 22.

Brick-work.

The dimensions of common bricks are from 7¾ to 8 inches long, by 4¼ wide, and 2¼ thick. Front bricks are 8¼ inches long, by 4½ wide, and 2½ thick.

The usual size of fire-bricks is 9⅛ inches long, by 4⅜ wide, by 2⅜ thick.

22 to 23 common bricks to a cubic foot when laid; 15 common bricks to a foot of 8-inch wall when laid.

To find the number of common bricks in a wall.

Rule.—Multiply the length of the wall in feet by the height in feet, and that by its thickness in feet, and that again by 22.

Example.—How many common bricks in a wall 40 feet long by 20 feet high, and 12 inches thick?

Solution.—40 feet length × 20 feet height, × 1 foot thick, × 22 = 17,600. *Ans.*

Note.—For walls 8 inches thick, multiply the length in feet by the height in feet, and that by 15.

When the wall is perforated by doors and windows, deduct the sum of their cubic feet from the cubic contents of the wall, including the openings, before multiplying by 22 or 15 as before.

Laths.

Laths are 1¼ to 1½ inches wide, by 4 feet long, are usually set ¼ inch apart, and a bundle contains 100.

BRICKLAYERS' WORK.

The principal is tiling, slating, walling and chimney work.

Of Tiling or Slating.

Tiling and slating are measured by the square of 100 feet, as flooring, partitioning and roofing were in the Carpenters' work; so that there is not much difference between the roofing and tiling; yet the tiling will be the most; for the bricklayers sometimes will require to have double measure for hips and valleys.

When gutters are allowed double measure, the way is to measure the length along the ridge-tile, and add it to the content of the roof: this makes an allowance of one foot in breadth, the whole length of the hips or valleys. It is usual also to allow double measure at the eaves, so much as the projector is over the plate, which is commonly about 18 or 20 inches.

Sky-lights and chimney shafts are generally deducted, if they be large, otherwise not.

Example 1. There is a roof covered with tiles, whose depth on both sides (with the usual allowance at the eaves) is 37 feet 3 inches, and the length 45 feet; how many squares of tiling are contained therein?

BY DUODECIMALS.		BY DECIMALS.
FEET.	INCHES.	37.25
37	3	45
45	0	
		18625
185		14900
148		
11	3	16 76.25
16 76	3	

2. *Of Walling.*

Bricklayers commonly measure their work by the rod of 16½ feet, or 272¼ square feet. In some places it is a custom to allow 18 feet to the rod; that is, 324 square feet. Sometimes the work is measured by the rod of 21 feet long and 3 feet high, that is, 63 square feet; and then no regard is paid to the thickness of the wall in measuring but the price is regulated according to the thickness.

When you measure a piece of brick-work, the first thing is to inquire by which of these ways it must be measured; then, having multiplied the length and breadth in feet together, divide the product by the proper divisor, viz.: 272.25, 324 or 63 according to the measure of the rod, and the quotient will be the answer in square rods of that measure.

But, commonly, brick walls that are measured by the rod are to be reduced to a standard thickness of a brick and a-half, which may be done by the following

RULE.—*Multiply the number of superficial feet that are contained in the wall by the number of half bricks which that wall is in thickness; one-third part of that product will be the content in feet.*

The dimensions of a building are generally taken by measuring half round the outside and half round the inside, for the whole length of the wall; this length, being multiplied by the hight, gives the superficies. And to reduce it to the standard thickness, etc., proceed as above. All the vacuities, such as doors, windows, window backs, etc., must be deducted.

To measure any arched way, arched window or door, etc., take the hight of the window or door from the crown or middle of the arch to the bottom or sill, and likewise from the bottom or sill to the spring of the arch; that is, where the arch begins to turn. Then to the latter hight add twice the former, and multiply the sum by the width of the window, door, etc., and one-third of the product will be the area, sufficiently near for practice

Example 1. If a wall be 72 feet 6 inches long, and 19 feet 3 inches high, and $5\frac{1}{2}$ bricks thick, how many rods of brick work are contained therein, when reduced to the standard?

GLAZIERS' WORK.

Glaziers take their dimensions in feet, inches and eights or tenths, or else in feet and hundredth parts of a foot, and estimate their work by the square foot.

Windows are sometimes measured by taking the dimensions of one pane, and multiplying its superficies by the number of panes. But, more generally, they measure the length and breadth of the window over all the panes and their frames for the length and breadth of the glazing.

Circular or oval windows, as fan lights, etc., are measured as if they were square, taking for their dimensions the greatest length and breadth, as a compensation for the waste of glass and labor in cutting it to the necessary forms.

Example 1. If a pane of glass be 4 feet $8\frac{3}{4}$ inches long, and 1 foot $4\frac{1}{4}$ inches broad, how many feet of glass are in that pane?

BY DUODECIMALS.	BY DECIMALS.
FT. IN. P.	4.729
4 8 9	1.354
1 4 3	-----
—————	18916
4 8 9	23645
1 6 11 0	14187
1 2 2 3	4729
—————————	—————
6 4 10 2 3	6.403066

Ans. 6 feet, 4 inches

PLUMBERS' WORK.

Plumbers' work is generally rated at so much per pound, or by the hundred weight of 112 pounds, and the price is regulated according to the value of lead at the time when the work is performed.

Sheet lead, used in roofing, guttering, etc., weighs from 6 to 12 pounds per square foot, according to the thickness, and leaden pipe varies in weight per yard, according to the diameter of its bore in inches.

The following table shows the weight of a square foot of sheet lead, according to its thickness, reckoned in parts of an inch, and the common weight of a yard of leaden pipe corresponding to the diameter of its bore in inches:

Thickness of Lead.	Pounds to a Square Foot.	Bore of Leaden Pipe.	Pounds per yard.
$\frac{1}{10}$	5.899	$\frac{3}{4}$	10
$\frac{1}{9}$	6.554	1	12
$\frac{1}{8}$	7.373	$1\frac{1}{4}$	16
$\frac{1}{7}$	8.427	$1\frac{1}{2}$	18
$\frac{1}{6}$	9.831	$1\frac{3}{4}$	21
$\frac{1}{5}$	11.797	2	24

Example 1. A piece of sheet lead measures 16 feet 9 inches in length, and 6 feet 6 inches in breadth; what is its weight at 8¼ pounds to a square foot?

BY DUODECIMALS			BY DECIMALS
FEET.	INCHES.		FEET.
16	9		16.75
6	6		6.5
100	6		8375
8	4	6	10050
108	10	6	108.875 *feet.*

Then 1 foot : 8¼ pounds :: 108.875 feet : 898.21875 pounds=8 cwt. 2¼ pounds nearly.

MASON'S WORK

Masons measure their work sometimes by the foot solid, sometimes by the foot superficial, and sometimes by the foot in length. In taking dimensions they girt all their moldings as joiners do.

The solids consist of blocks of marble, stone pillars, columns, etc. The superficies are pavements, slabs, chimney-pieces, etc.

PLASTERERS' WORK.

Plasterers' work is principally of two kinds; namely, plastering upon laths, called *ceiling*, and plastering upon walls or partitions made of framed timber, called *rendering*.

In plastering upon walls, no deductions are made except for doors and windows, because cornices, festoons, enriched moldings, etc., are put on after the room is plastered.

In plastering timber partitions, in large warehouses, etc., where several of the braces and larger timbers project from the plastering, a fifth part is commonly deducted. Plastering between their timbers is generally called rendering between quarters.

Whitening and coloring are measured in the same manner as plastering; and in timbered partitions, one-fourth, or one-fifth of the whole area is commonly added, for the trouble of coloring the sides of the quarters and braces.

Plasterers' work is measured by the yard square, consisting of nine square feet. In arches, the girt round them, multiplied by the length, will give the superficies.

Example 1.—If a ceiling be 59 feet 6 inches long, and 24 feet 6 inches broad, how many yards does that ceiling contain?

PROBLEM I.

To find the solid content of a Dome, having the hight and the dimensions of its base given.

Rule.—*Multiply the area of the base by the hight, and ⅔ of the product will be the solidity.*

Example 1.—What is the solidity of a dome, in the form of a hemisphere, the diameter of the circular base being 60 feet?

$60^2 \times .7854 = 2827.44$ area of the base.

Then ⅔ $(2827.44 \times 30) = 56548.8$ cubic feet. *Ans.*

PROBLEM II.

To find the superficies of a dome, having the hight and dimensions of its base given.

Rule.—*Multiply the area of the base by 2, and the product will be the superficial content required; or, multiply the square of the diameter of the base by 1.5708.*

For an Elliptical Dome.—*Multiply the two diameters of the base together, and that product by 1.5708, the last product will be the area, sufficiently correct for practical purposes.*

SHORT RULES FOR THE MECHANIC.

QUESTION.—A stick of timber is carried by three men, one carries at the end, and the other two with a lever, How far should the lever be placed from the other end, that *each* man may carry equally?

RULE.—Divide the length of the stick by 4, and the quotient is the answer.

There is a stick of timber, 30 feet long, to be carried by 3 men: one carries at the end, the other two carry by a lever; how far must the *lever* be placed from the other end, that each may carry equally? *Ans.* 7½ feet from the end.

SQUARE & CUBE ROOTS

To work the square and cube roots with ease and facility, the pupil must be familiar with the following properties of numbers:

Their importance can not be exaggerated if we wish to insure skill or even sound information on this subject.

I. A square number, multiplied by a square number, the product will be a square number.

II. A square number, divided by a square number, the quotient is a square.

III. A cube number, multiplied by a cube, the product is a cube.

IV. A cube number, divided by a cube, the quotient will be a cube.

V. If the square root of a number is a composite number, the square itself *may be divided into integer square factors*; but if the root is a *prime number*, the square can not be separated into square factors *without fractions*.

VI. If the unit figure of a square number is 5, we may multiply by the square number 4, and we shall have another square, whose *unit* period will be ciphers.

VII. If the unit figure of a cube is 5, we may multiply by the cube number 8, and produce another cube, whose unit period will be ciphers.

N. B. If a supposed cube, whose unit figure is 5, be multiplied by 8, and the product does not give three ciphers on the right, the number *is not a cube.*

We present the following table, for the pupil to compare the natural numbers with the *unit figure* of their *squares* and *cubes*, that he may be able to *extract roots by inspection.*

Numbers	1	2	3	4	5	6	7	8	9	10
Squares	1	4	9	16	25	36	49	64	81	100
Cubes	1	8	27	64	125	216	343	512	729	1000

EXERCISES FOR PRACTICE.

1. What is the square root of 625? *Ans.* 25.

If the root is an *integer number*, we may know, by the inspection of the table, that it must be 25, as the greatest square in 6 is 2, and 5 is the only figure whose square is 5 in its unit place.

Again, take 625
Multiply by 4   4 being a square.
       2500

The square root of this product is obviously 50, but this must be divided by 2, the square root of 4, which gives 25, the root.

2. What is the square root of 6561? *Ans.* 81.

As the *unit figure* in this example is 1, and in

the line of squares in the table, we find 1 only at 1 and 81, we will, therefore, divide 6561 by 81, and we find the quotient 81; 81 is, therefore, the square root.

3. What is the square root of 106729? *Ans.* 327

As the unit figure, in this example, is 9, if the number is a square, it must divide by either 9, or 49. After dividing by 9 we have 11881 for the other factor, a prime number, therefore its root is a prime number=109. 109, multiplied by 3, the root of 9, gives 327 for the answer.

4. What is the root of 451584? *Ans.* 672.

As the unit figure is 4, and in the line of squares we find 4 only at 4 and 64, the above number, *if a square*, must divide by 4, or 64, or by both.

We will divide it by 4, and we have the factors 4 and 112896. This last factor *closes in* 6; therefore, by looking at the table, we see it must divide by 16, or 36, etc.

We divide by 36, and we have the factors 36 and 3136; divide this last by 16, and we have 16 and 196; divide this last fraction by 4, and we have 4 and 49.

Take now our divisors, and last factor, 49, and we have for the original number the product of $4 \times 36 \times 16 \times 4 \times 49$; the roots of which are $2 \times 6 \times 4 \times 2 \times 7$, the products of which are 672, the answer

5. Extract the square root of 2025. *Ans.* 45.

1st. Divide by the square number 25, and we find the two factors, 25×81, as equivalent to the given number. Roots of these factors, 5×9=45, the answer.

Again, multiply by the square number 4, when a number ends in 25, and we have 8100, root 90, half of which, because we multiplied by 4, the square of 2, is 45, the answer.

Problems on the Right-angled Triangle.

1. The top of a castle is 45 yards high, and is surrounded with a ditch, 60 yards wide; required the length of a ladder that will reach from the outside of the ditch to the top of the castle.

Ans. 75 yards.

This is almost invariably done by squaring 45 and 60, adding them together, and extracting the square root; but so much labor *is never necessary when the numbers have a common divisor*, or when the side sought is expressed by a *composite* number.

Take 45 and 60; both may be divided 15, and they will be reduced to 3 and 4. Square these, 9+16=25. The square root of 25 is 5, which multiplied by 15, gives 75, the answer.

Abbreviations in Cube Root

1. What is the cube root of 91125? *Ans.* 45.

$$\begin{array}{r} \text{Multiply by} \quad 8 \\ \hline 729000 \end{array}$$

Now, 729 being the cube of 9, the root of 729000 is 90; divide this by 2, the cube root of 8. and we have 45, the answer.

When it is requisite to multiply several numbers together and extract the cube root of their product, try to change them into *cube factors* and extract the root *before* multiplication.

EXAMPLES.

1. What is the side of a cubical mound equal to one 288 feet long, 216 feet broad, and 48 feet high?

The common way of doing this, is to multiply these numbers together and extract the root—a lengthy operation. But observe that 216 is a cube number, and $288 = 2 \times 12 \times 12$, and $48 = 4 \times 12$; therefore the whole product is $216 \times 8 \times 12 \times 12 \times 12$. Now, the cube root of 216 is 6, of 8 is 2, and of 12^3 is 12, and the product of $6 \times 2 \times 12 = 144$, the answer.

2. Required the cube root of the product of 448×392 the short way.   *Ans.* 56.

We can extract the root of cube numbers by inspection when they do not contain more than two periods.

SQUARE AND CUBE ROOTS.

RULE.—*As there will be two figures in the root, the first may easily be found mentally, or by the Table of Powers; and if the unit figure of the power be* 1, *the unit figure in the root will be* 1; *and if it be* 8, *the root will be* 2; *and if* 7, *it will be* 3; *and if the unit of the power be* 6, *the unit of the root will be* 6; *and if* 5, *it will be* 5; *if* 3, *it will be* 7; *if* 2, *it will be* 8; *and if the unit of the power be* 9, *the unit of the root will be* 9. *This will appear evident by inspecting the Table of Powers.*

EXAMPLES.

Find the cube root of 195112. This number consists of two periods. Compare the superior period with the cubes in the table, and we find that 195 lies between 125 and 216. The cube root of the tens, then, must be 5. The unit figure of the given cube is 2; and no cube in the table has 2 for its unit figure, except 512, whose root is 8; therefore 58 is the root required.

What is the cube root of 97336?   *Ans* 46.

EXPLANATION.—By examining the left hand period, we find the root of 97 is 4, and the cube of 4 is 64. The root can not be 5, because the cube of 5 is 125. The unit figure of the given cube is 6; and no cube in the table has 6 for its unit figure, except 216, whose root is 6; the answer, therefore, is 46.

The number 912673 is a cube, what is its root?
Ans. 97.

Observe, the root of the superior period must

be 9, and the root of the unit period must be some number which will give 3 for its unit figure *when cubed*; and 7 is the only figure that will answer.

The following numbers are cubes; required their roots

1. What is the cube root of 59319?   *Ans.* 39
2. What is the cube root of 79507?   *Ans.* 43.
3. What is the cube root of 117649?   *Ans.* 49.
4. What is the cube root of 110592?   *Ans.* 48.
5. What is the cube root of 357911?   *Ans.* 71.
6. What is the cube root of 389017?   *Ans.* 73.
7. What is the cube root of 571787?   *Ans.* 83.

When a cube has more than two periods, it can generally be reduced to two by dividing by some one or more of the cube numbers, unless the root is a *prime* number.

The number 4741632 is a cube; required its root. Here we observe that the unit figure is 2; the unit figure of the root must therefore be the root of 512, as that is the only cube of the 9 digits whose unit figure is 2. The cube root of 512 is 8; therefore 8 is the unit figure in the root, and the root is an *even* number, and can be divided by 2, and of course the cube itself can be divided by 8, the cube of 2.   8)4741632
                                    ─────────
                                     592704

Now as the first number was a cube, and being

divided by a cube, the number 592704 must be a cube, and by inspection, as previously explained, its root must be 84, which, multiplied by 2, gives 168, the root required.

The number 13312053 is a cube; what is its root? *Ans.* 237.

As there are three periods, there must be three figures, units, tens, and hundreds, in the root; the hundreds must be 2, the units must be 7. Let us then divide the 2d figure, or the tens, *in the usual way*, and we have 237 for the root.

Again, divide 13312053 by 27, and we have 493039 for another factor. The root of this last number must be 79, which, multiplied by 3, the cube root of 27, gives 237, as before.

The number 18609625 is a cube; what is its root?

As this cube ends with 5, we will multiply it by 8:

$$\begin{array}{r} 18609625 \\ 8 \\ \hline 148877000 \end{array}$$

As the first is a cube, this product must be a cube; and as far as labor is concerned, it is the same as reduced to two periods, and the root, we perceive at once, must be 530, which, divided by 2, gives 265 for the root required.

N. B.—If a number, whose unit figure is 5, be multiplied by 8, and *does not result in three ciphers* on the right, the number is not a cube.

To find the Approximate Cube Root of Surds.

RULE.—*Take the nearest rational cube to the given number, and call it the assumed cube; or assume a root to the given number and cube it. Double the assumed cube and add the number to it; also double the number and add the assumed cube to it. Take the difference of these sums, then say, As double of the assumed cube, added to the number, is to this difference, so is the assumed root to a correction.*

This correction, added to or subtracted from the assumed root, *as the case may require*, will give the cube root very nearly.

By repeating the operation with the root last found as an assumed root, we may obtain results to any degree of exactness; one operation, however, is generally sufficient.

EXAMPLES.

1. Required the cube root of 66.

The cube root of 64 is 4. Now it is manifest that the cube root of 66 is a little more than 4, and by taking a similar proportion to the preceding, we have

$64 \times 2 = 128$ $2 \times 66 = 132$
    66   64
   ———   ———
   194   196 : : 4 : to root of 66.

Or, 194 : 2 :: 4 : to a correction

```
194)8.0000(0.04124
     7 76
     ———
       240
       194
       ———
        460
        388
        ———
         720
```

Therefore the cube root of 66 is 4.04124.

2. Required the cube root of 123.

Suppose it 5; cube it, and we have 125.

Now we perceive that the cube of 5 being greater than 123, the *correction* for 5 must be *subtracted.*

```
      2×125=250   246
Add          123  125
             ———  ———
As           373 : 371 :: 5 : root of 123.
Or,  373 : 2 :: 5 : correction for 5
     373)10.0000(0.02681
          7 46
          ———
          2 540       From 5.00000
          2 238       Take 0.02681
          ———         ———
           3020   Ans. 4.97319
           2984
           ———
            360.
```

3. What is the cube root of 28? *Ans.* 3,03658+
4. What is the cube root of 26? *Ans.* 2,96249+
5. What is the cube root of 214? *Ans.* 5,98142+
6. What is the cube root of 346? *Ans.* 9,02034+

The above being very near integral cubes— that is, 28 and 26 are both near the cube number 27, 214 is near 216, etc. All numbers very near cube numbers are *easy of solution*.

We now give other examples, more distant from integral cubes, to show that the labor must be more lengthy and tedious, though the operation is the same.

1. What is the cube root of 3214? *Ans.* 14,75758.

Suppose the root is 15—its cube is 3375, which, being greater than 3214, shows that 15 is too great; the correction will therefore be subtractive.

By the rule, 9964 : 161 :: 15. 0,243, the correction.

Assumed root..........................15,0000
Less.................................... 2423

Root nearly......................... 14,7577

Now assume 14,7 for the root, and go over the operation again, and you will have the true root to 8 or 10 places of decimals.

N. B.—Roots of component powers may be obtained more readily thus :

For the 4th root, take the square root of the square root.

MENSURATION OR PRACTICAL GEOMETRY.

MEASUREMENT OF GRINDSTONES.

Grindstones are sold by the stone, and their contents found as follows: *

RULE.—*To the whole diameter add half of the diameter, and multiply the sum of these by the same half, and this product by the thickness; divide this last number by 1728, and the quotient is the contents, or answer required.*

EXAMPLES.

What are the contents of a grindstone 24 inches diameter, and 4 inches thick

$$\frac{24+12 \times 12 \times 4}{1728} = 1 \text{ stone. } Ans.$$

2. What are the contents of a grindstone 36 inches diameter, and 4 inches thick. *Ans.* $2\frac{1}{4}$ stone.

Mensuration of Superficies and Solids.

Superficial measure is that which relates to length and breadth only, not regarding thickness. It is made up of squares, either greater or less, according to the different measures by which the dimensions of the figure are taken or measured. Land is measured by this measure, its dimensions being

---

*24 inches in diameter, and 4 inches thick make a stone

usually taken in acres, rods, and links. The contents of boards, also, are found by this measure, their dimensions being taken in feet and inches. Because 12 inches in length make 1 foot of long measure, therefore 12×12=144, the square inches in a superficial foot, etc.

NOTE.—Superficial means lying on the surface

To find the area of a square having equal sides.

RULE.—*Multiply the side of the square into itself, and the product will be the area, or superficial content of the same name with the denomination taken, whether inches, feet, yards, rods, and links, or acres.*

EXAMPLES.

1. How many square feet of boards are contained in the floor of a room which is 20 feet square?

20×20=400 feet, the answer.

2. Suppose a square lot of land measures 36 rods on each side, how many acres does it contain?

36×36=1296 square rods. And 1296÷160=8 acres, 16 rods, *Ans.*

As 160 square rods make an acre, therefore we divide 1296 by 160 to reduce rods to acres.

N. B.—The shortest way to work this example is, to cancel 36×36 with the divisor 160. Arrange the example as below; (divide both terms by 4×4:)

$$\frac{36 \times 36}{160} \text{ same as } \frac{9 \times 9}{10} = 8.1 \text{ acres, or 8ac. 16 rods.}$$

MENSURATION OR PRACTICAL GEOMETRY. 239

To measure a parallelogram or long square.

RULE.—*Multiply the length by the breadth, and the product will be the area, or superficial contents in the same name as that in which the dimension was taken, whether inches, feet, or rods, etc.*

EXAMPLES

1. A certain garden, in form of a long square, is 96 feet long, and 54 feet wide; how many square feet of ground are contained in it?

Ans. $96 \times 54 = 5184$ square feet.

2. A lot of land, in form of a long square, is 120 rods in length, and 60 rods wide; how many acres are in it? $120 \times 60 = 7200$ sq. rods. And $7200 \div 160 = 45$ acres, *Ans.*

NOTE.—The learner must recollect that feet in length, multipled by feet in breadth, produce *square feet*; and the same of the other denominations of lineal measure.

NOTE.—Both the length and breadth, if not in units of the same denomination, must be made so before multiplying.

3. How many acres are in a field of oblong form, whose length is 14,5 chains, and breath 9,75 chains? Ans. 14ac. 0rood, 22rods.

NOTE.—The Gunter's chain is 66 feet, or 4 rods, long, and contains 100 links. Therefore if dimensions be given in chains and decimals, point off from the product one more decimal place than are contained in both factors, and it will be acres and decimals of an acre; if in chains and links, do the same, because links are hundredths of chains, and therefore the same as decimals of them. Or, as 1 chain wide, and 10 chains long, or 10 square chains, or 100000 square links, make an acre, it is the same as if you divide the links in the area by 100000.

4. If a board be 21 feet long and 18 inches broad, how many square feet are contained in it?

18 inches=1,5 foot; and $21 \times 1,5 = 31,5$ ft., *Ans.*

Or, in measuring boards, you may multiply the length in feet by the breadth in inches, and divide the product by 12; the quotient will give the answer in square feet, etc.

Thus, in the last example, $\dfrac{21 \times 18}{12} = 31\tfrac{1}{2}$ sq. ft., as before.

5. If a board be 8 inches wide, how much in length will make a foot square?

RULE.—*Divide 144 by the width; thus,* 8)144

Ans. 18 *in.*

6. If a piece of land be 5 rods wide, how many rods in length will make an acre?

RULE.—*Divide 160 by the width, and the quotient will be the length required, thus,*
5)160
—————
Ans. 32 *rods in length.*

MENSURATION OR PRACTICAL GEOMETRY.

NOTE.—When a board, or any other surface is wider at one end than the other, but yet is of a true taper, you may take the breadth in the middle or add the widths of both ends together, and halve the sum for the mean width; then multiply the said mean breadth in either case by the length; the product is the answer or area sought.

7. How many square feet in a board, 10 feet long and 13 inches wide at one end, and 9 inches wide at the other?

$$\frac{13+9}{2} = 11 \text{ in., mean width.}$$

$$\frac{10 \times 11}{12} = 9\tfrac{1}{6}\text{ft., } Ans.$$

8. How many acres are in a lot of land which is 40 rods long, and 30 rods wide at one end, and 20 rods wide at the other?

$$\frac{30+20}{2} = 25 \text{ rods, mean width.}$$

Then, $$\frac{25 \times 40}{160} = 6\tfrac{1}{4} \text{ acres, } Ans.$$

9. If a farm lie 250 rods on the road, and at one end be 75 rods wide, and at the other 55 rods wide, how many acres does it contain?

Ans. 101 acres, 2 roods, 10 rods.

N. B.—Always arrange your example as above, and cancel the factors common to both terms before multiplying

CASE 3.—*To measure the surface of a triangle.*

DEFINITION.—A triangle is any three-cornered figure which is bounded by three right lines.*

RULE.—*Multiply the base of the given triangle into half its perpendicular hight, or half the base into the whole perpendicular, and the product will be the area.*

EXAMPLES.

1. Required the area of a triangle whose base or longest side is 32 inches, and the perpendicular hight 14 inches.

$14 \div 2 = 7 =$ half the perpendicular. And $32 \times 7 = 224$ sq. in., *Ans.*

2. There is a triangular or three-cornered lot of land whose base or longest side is $51\frac{1}{2}$ rods; the perpendicular, from the corner opposite to the base, measures 44 rods; how many acres does it contain?

$44 \div 2 = 22 =$ half the perpendicular.

$$\frac{\text{And } 51.5 \times 22}{160} = 7 \text{ acres, 13 rods. } Ans.$$

Joists and planks are measured by the following

RULE.—*Find the area of one side of the joist or plank by one of the preceding rules; then multiply it by the thickness in inches, and the last product will be the superficial content.*

---

*A triangle may be either right-angled or oblique.

MENSURATION OR PRACTICAL GEOMETRY. 243

EXAMPLES.

1. What is the area, or superficial content, or board measure, of a joist, 20 feet long, 4 inches wide, and 3 inches thick? $\dfrac{20 \times 4}{12} \times 3 = 20$ ft., *Ans*

2. If a plank be 32 feet long, 17 inches wide, and 3 inches thick, what is the board measure of it? *Ans.* 136 feet

NOTE.—There are some numbers, the sum of whose squares makes a perfect square; such are 3 and 4, the sum of whose squares is 25, the square root of which is 5; consequently, when one leg of a right-angled triangle is 3, and the other 4, the hypotenuse must be 5. And if 3, 4, and 5, be multiplied by any other numbers, each by the same, the products will be sides of true right-angled triangles. Multiplying them by 2, gives 6, 8, and 10, by 3, gives 9, 12, and 15; by 4, gives 12, 16, and 20, etc.; all which are sides of right-angled triangles. Hence architects, in setting off the corners of buildings, commonly measure 6 feet on one side, and 8 feet on the other; then, laying a 10-foot pole across from those two points, it makes the corner a true right-angle.

N. B.—The solutions of the foregoing problems are all very brief by canceling.

To find the area of any triangle when the three sides only are given.

RULE.—*From half the sum of the three sides subtract each side severally; multiply these three remainders and the said half sum continually together; then the square root of the last product will be the area of the triangle.*

EXAMPLE.

Suppose I have a triangular fish-pond, whose three sides measure 400, 348, and 312yds; what quantity of ground does it cover?

Ans. 10 acres, 3 roods, 8+rods.

NOTE.—If a stick of timber be hewn three square, and be equal from end to end, you find the area of the base, as in the last question, in inches; multiply that area by the whole length, and divide the product by 144, to obtain the solid content.

If a stick of timber be hewn three square, be 12 feet long, and each side of the base 10 inches, the perpendicular of the base being $8\frac{2}{3}$ inches, what is its solidity? *Ans.* 3,6+feet.

PROBLEM 1.

The diameter given, to find the circumference.

RULE.—*As 7 are to 22, so is the given diameter to the circumference; or, more exactly, as 113 are to 355, so is the diameter to the circumference, etc*

MENSURATION OR PRACTICAL GEOMETRY. 245

EXAMPLES.

1. What is the circumference of a wheel, whose diameter is 4 feet?

As 7 : 22 : : 4 : 12,57+ft., the circum., *Ans.*

2. What is the circumference of a circle, whose liameter is 35 rods?

As 7 : 22 : : 35 : 110 rods, *Ans.*

NOTE.—To find the diameter when the circumference is given, reverse the foregoing rule, and say, as 22 are to 7, so is the given circumference to the required diameter; or, as 355 are to 113, so is the circumference to the diameter.

3. What is the diameter of a circle, whose circumference is 110 rods?

As 22 : 7 : : 110 : 35 rods, the diam., *Ans.*

CASE 5.—*To find how many solid feet a round stick of timber, equally thick from end to end, will contain, when hewn square.*

RULE.—*Multiply twice the square of its semi-diameter, in inches, by the length in the feet; then divide the product by 144, and the quotient will be the answer.*

N. B.—When multiplication and division are combined, always cancel like factors. When the numbers are properly arranged, a few clips with the pencil, and, perhaps, a *trifling* multiplication will suffice.

For the practical convenience of those who have occasion to refer to mensuration, we have arranged the following useful table of multiples. It covers the whole ground of practical geometry, and should be studied carefully by those who wish to be skilled in this beautiful branch of mathematics:

TABLE OF MULTIPLES.

Diameter of a circle × 3.1416 = Circumference.
Radius of a circle × 6.283185 = Circumference.
Square of the radius of a circle × 3.1416 = Area.
Square of the diameter of a circle × 0.7854 = Area.
Square of the circumference of a circle × 0.07958 = Area.
Half the circumference of a circle × by half its diameter = Area.
Circumference of a circle × 0.159155 = Radius.
Square root of the area of a circle × 0.56419 = Radius.
Circumference of a circle × 0.31831 = Diameter.
Square root of the area of a circle × 1.12838 = Diameter.
Diameter of a circle × 0.86 = Side of inscribed equilateral triangle.
Diameter of a circle × 0.7071 = Side of an inscribed square.
Circumference of a circle × 0.225 = Side of an inscribed square.
Circumference of a circle × 0.282 = Side of an equal square.
Diameter of a circle × 0.8862 = Side of an equal square.
Base of a triangle × by ½ the altitude = Area.
Multiplying both diameters and .7854 together = Area of an ellipse.
Surface of a sphere × by ⅙ of its diameter = Solidity.
Circumference of a sphere × by its diameter = Surface.
Square of the diameter of a sphere × 3.1416 = Surface.
Square of the circumference of a sphere × 0.3183 = Surface.
Cube of the diameter of a sphere × 0.5236 = Solidity.
Cube of the radius of a sphere × 4.1888 = Solidity.
Cube of the circumference of a sphere × 0.016887 = Solidity.
Square root of the surface of a sphere × 0.56419 = Diameter.
Square root of the surface of a sphere × 1.772454 = Circumference.
Cube root of the solidity of a sphere × 1.2407 = Diameter.
Cube root of the solidity of a sphere × 3.8978 = Circumference.
Radius of a sphere × 1.1547 = Side of inscribed cube.
Square root of (⅓ of the square of) the diameter of a sphere = Side of inscribed cube.
Area of its base × by ⅓ of its altitude = Solidity of a cone or pyramid, whether round, square, or triangular.
Area of one of its sides × 6 = Surface of a cube.
Altitude of trapezoid × ½ the sum of its parallel sides = Area.

AVOIRDUPOIS WEIGHT

Is the standard weight for weighing the greater portion of articles used in trade and commerce, such as groceries, produce, iron, coal, hay, cotton, etc.

TABLE.

437½ grains (*gr*).1 ounce....*oz*.
16 oz........1 pound....*lb*.
25 lb........1 quarter...*qr*.
4 qr........1 h'd.w'ht,*cwt*.
20 cwt.......1 ton.......*T*.

EQUIVALENTS.

T.	*cwt.*	*qr.*	*lb.*	*oz.*	*gr.*
1 =	20 =	80 =	2000 =	32000 =	14000000
	1 =	4 =	100 =	1600 =	700000
		1 =	25 =	400 =	175000
			1 =	16 =	7000
				1 =	437½

Scale of units :—437½, 16, 25, 4, 20.

The dram is now seldom used, except with silk manufacturers. The ounce is divided into ¼ and ½.

The following denominations are also used·

LONG OR IRON TON.

28 lbs..1 quarter.
4 qr., or 112 lbs..........................1 hundredweight.
20 cwt., or 2240 lbs......................1 ton.

This measurement is nearly obsolete. It is allowed at the Custom House in estimating duties, and in wholesale transactions of iron and coal.

NOTE.—The grain avoirdupois, though never used, is the same as the grain in Troy weight. 7000 grains make the avoirdupois pound, and 5760 grains the Troy pound.

IRON, LEAD, Etc.

14 lbs...1 stone.
21½ stone..1 pig.
8 pigs..1 fother.

AVOIRDUPOIS WEIGHT.
Miscellaneous Table.

14 pounds	of	Iron or Lead............	1 stone.
100	"	" Grain or Flour...........	1 cental.
100	"	" Raisins	1 cask.
100	"	" Dry Fish	1 quintal.
100	"	" Nails	1 keg.
196	"	" Flour	1 barrel.
200	"	" Pork, Beef, or Fish......	1 barrel.
240	"	" Lime	1 cask.
280	"	" Salt	1 barrel of Salt at the N.Y. Salt Works.

APOTHECARIES' WEIGHT

Is used by Apothecaries and Physicians in dispensing medicines, not liquid. The grains mentioned in the following table are *Troy.*

TABLE.

20 Grains (*gr.* xx.)............1 scruple, ℈
3 Scruples (℈ iij)............1 dram, ℨ
8 Drams (ℨ viij)............1 ounce, ℥
12 Ounces (℥ xij)............1 pound, ℔

EQUIVALENTS.

℔		℥		ℨ		℈		gr.
1	=	12	=	96	=	288	=	5760
		1	=	8	=	24	=	480
				1	=	3	=	60
						1	=	20

SCALE OF UNITS :—20, 3, 8, 12.

The only difference between *Troy* and *Apothecaries'* weight is the division of the ounce. The pound, ounce, and grain are the same. Drugs and medicines are bought and sold in quantities by Avoirdupois weight.

APOTHECARIES' FLUID MEASURE,

Used in mixing liquid medicines by measure.

```
60 Minims (m)......1 fluid drachm, fʒ
 8 fʒ..............1 fluid ounce,  f℥
16 f℥..............1 pint,   O. (Octarius.)
 8 O..............1 gallon, (Cong. Congius.)
```

EQUIVALENTS.

Cong.	O.	fʒ	fʒ	m
1 =	8 =	128 =	1024 =	61440
	1 =	16 =	128 =	7680
		1 =	8 =	480
			1 =	60

SCALE OF UNITS :—60, 8, 16, 8.

NOTE.—One fluid ounce=455.6944 Troy grains.

The *minim* is a drop of pure water, and is equal to about $\frac{95}{100}$ of a grain Troy.

An ordinary teacupful is about 4 *fluid ounces.*
Common tablespoonful ½ a *fluid ounce.*
Teaspoon contains about 45 drops.

WOOD MEASURE,

For measuring wood, rough stone, fences, etc.

```
16 Cu. Ft..........1 cord ft.
 8 Cord Feet or
128 Cubic Ft..1 cord.
24¾ Cubic Feet......1 perch
    of stone or masonry.
```

A cord of wood is a pile 8 feet long, by 4 feet wide, and 4 feet high.

A cord foot is one foot of the running pile, or ⅛ of a cord.

A perch of stone or masonry is 16½ feet long, 1½ feet wide, and 1 foot high.

Wood. Measuring wood in the load. If the rack is narrower at the bottom than at the top, the width of the load should be measured half-way from base to top; this will give the average width.

DRY MEASURE

Is used in measuring articles not fluid, such as grains, seeds, vegetables, fruit, salt, etc.

TABLE.

2 pints(*pt.*)..1 quart.... *qt.*
8 qt............1 peck......*pk.*
4 pk...........1 bushel....*bu.*
36 bu ....1 ch'ldron,*ch.*

EQUIVALENTS.

ch.		*bu.*		*pk.*		*qt.*		*pt.*
1	=	36	=	144	=	1152	=	2304
		1	=	4	=	32	=	64
				1	=	8	=	16
						1	=	2

Scale of units :—2, 8, 4, 36.

NOTE.—1 gal. Wine Measure contains 231 cu. in., 1 gal. Ale or Beer Measure (nearly obsolete), 282 cu. in., and 1 bu. $2150\tfrac{42}{100}$ cu. in.

The legal bushel of the United States is the old Winchester measure, cylindrical in form, 18½ inches in diameter and 8 inches deep, containing $2150\tfrac{42}{100}$ cubic inches. The Imperial bushel of England is $2218\tfrac{142}{1000}$ cubic inches. 32 English bushels equal 33 of the United States.

Heaped Measure is the contents of bushel heaped in the shape of a cone. Corn in the ear, large fruits, vegetables, and bulky articles are sold by this measure.

Stricken Measure is the bushel even full, having been stricken off by a rule or striker. Grain, seeds, etc., are sold by this measure. It is customary to allow 5 stricken measures for 4 heaped ones. It is usual to quote the price of grain, etc., by the bushel, but more frequently to determine their value by weight.

CUBIC OR SOLID MEASURE,

Used in measuring anything containing length, breadth, depth, and thickness, such as timber, wood, stone, boxes, storage capacity of rooms, bins, cisterns, etc.

EQUIVALENTS.

cu. ft.   cu. in.
1   =   1728

SCALE OF UNITS:—1, 1728.

TABLE.

1728 Cubic Inches (*cu. in.*)......1 cubic foot, *cu. ft.*
27 Cubic Feet......................1 cubic yard, *cu. yd.*
40 Cubic Feet.....................1 ton of ship cargo.
50 Cu. Ft. of Square Timber..1 ton.

One cubic yard contains 46,656 cubic inches.

A Registered ton, in computing the tonnage of ships and vessels, is 100 cubic feet of internal capacity.

In measuring cargoes, a ton is 40 cubic feet in the United States, and 40 cubic feet in England.

Light articles of freight are generally estimated by the space occupied, but heavy articles by weight.

A CUBIC FOOT OF

	Pounds.			Pounds.
Common soil	weighs 124	Clay and stones	weigh	160
Strong "	" 127	Cork	weighs	15
Loose earth or sand "	95	Tallow	"	59
Clay	" 135	Bricks	"	125
Lead	" 708¾	Marble	"	171
Brass	" 534¾	Granite	"	165
Copper	" 555	Sea-water	"	64 3/16
Wrought iron	" 486¾	Oak wood	"	55
Anthracite coal	" 50–55	Red pine	"	42
Bituminous "	" 45–55	White pine	"	30
Charcoal (hard w'd) "	18½	Charc'l (pine wood) "		18

MEASUREMENTS AND ESTIMATES.

A cubic yard of earth is called a load.

Bricks are of various dimensions. The average size is 8 inches long, 4 inches wide, 2 inches thick. 27 bricks make a cubic foot, when laid dry. Laid in mortar $\frac{1}{8}$ to $\frac{1}{16}$ is allowed for mortar. Baltimore and Milwaukee bricks are $8\frac{1}{4} \times 4\frac{1}{2} \times 2\frac{3}{8}$ inches.

Brick-Work is generally estimated by the 1000. When measured by square measure the work is understood to be 12 inches thick.

Board and Lumber Measure. All estimates are made on one inch in thickness; for every $\frac{1}{4}$ inch in thickness $\frac{1}{4}$ price is added.

Board feet are changed to cubic feet by dividing by 12.
Cubic feet to board, by multiplying by 12.

Estimating Work by Artificers.

In material only is allowance made for windows, doors and cornices. No allowance being made in estimating the work. The size of a cellar or wall is estimated by the measurement of the outside. No allowance for corners.

Estimates, How Made.

By the square foot, as in glazing, stone-cutting, etc.
By the square yard, as in plastering, painting, etc.
By the square (100 sq. ft.), as in flooring, roofing, slating, paving, etc.
Painting of mouldings, cornices, etc., the estimate is by measuring the entire surface.

SIZE OF NAILS.

2-penny.........1	inch.........557	nails per pound.
4-penny.........1$\frac{1}{4}$	inches......353	"	"
5-penny.........1$\frac{3}{4}$	"	232	"	"
6-penny 2	"	167	"	"
7-penny.........2$\frac{1}{4}$	"	141	"	"
8-penny.........2$\frac{1}{2}$	"	101	"	"
10-penny........2$\frac{3}{4}$	"	 68	"	"
12-penny........3	"	 54	"	"
20-penny........3$\frac{1}{2}$	"	 34	"	"
Spikes.........4	"	 16	"	"
Spikes.........4$\frac{1}{2}$	"	 12	"	"
Spikes.........5	"	 10	"	"

From this table an estimate of quantity and suitable sizes for any job of work can be made.

LIQUID OR WINE MEASURE.

Used in measuring liquids, such as liquors, vinegar, molasses, oils, etc., and estimating the capacity of vessels designed to contain them.

TABLE.

4 gills (*gi.*) 1 pint, *pt.*
2 pints ......1 quart, *qt.*
4 quarts ....1 gallon, *gal.*
31½ gallons...1 barrel, *bbl.*
2 barrels...1 hogs'd, *hhd.*

EQUIVALENTS.

hhd.		bbl.		gal.		qt.		pt.		gi.
1	=	2	=	63	=	252	=	504	=	2016
		1	=	31½	=	126	=	252	=	1008
				1	=	4	=	8	=	32
						1	=	2	=	8
								1	=	4

SCALE OF UNITS:—4, 2, 4, 31½, 2.

NOTE.—The gallon must contain exactly 10 pounds troy, of pure water, at a temperature of 62°, the barometer being at 30 inches. It is the standard unit of measure of capacity for liquids and dry goods of every description, and is $\frac{1}{5}$ larger than the old wine measure, $\frac{1}{32}$ larger than the old dry measure, and $\frac{1}{80}$ less than the old ale measure. The wine gallon must contain 231 cubic inches.

Barrels used in commerce are made of various sizes, from 30 to 45, and even 56 gallons. There is no definite measure called a hogshead, they are usually gauged, and have their capacities in gallons marked on them. The Standard gallon, United States, contains 231 cubic inches. The Imperial gallon, Great Britain, 277.274 cubic inches, and is equal to $\frac{1}{5}$ more than the United States' measure.

In measuring cisterns, reservoirs, vats, etc., the barrel is estimated at 31½ gallons, and the hogshead 63 gallons.

A gallon of water weighs nearly 8½ pounds, avoirdupois.

A pint is generally estimated as a pound.

LINEAR OR LONG MEASURE.

For measuring lengths, distances and dimensions of objects.

TABLE.

12 inches.............1 foot.
3 feet................1 yard.
5½ yards..............1 rod.
16½ feet..............1 rod.
320 rods..............1 mile.

NOTE.—The inch is divided into ½, ¼, ⅛.

EQUIVALENTS.

mi.		fur.		rd.		yd.		ft.		in.
1	=	8	=	320	=	1760	=	5280	=	63360
		1	=	40	=	220	=	660	=	7920
				1	=	5½	=	16½	=	198
						1	=	3	=	36
								1	=	12

Scale of units:—12, 3, 5½, 40, 8.

NOTE.—*Cloth Measure* is practically out of use. In measuring goods sold by the yard, the yard is divided into *halves, fourths, eighths,* and *sixteenths.* At United States Custom Houses, in estimating duties, the yard is divided into *tenths* and *hundredths.*

FOR MEASURING HEIGHTS AND DISTANCES.

3 inches.............1 palm. | 9 inches..............1 span.
4   "   ....1 hand. | $3\frac{3}{10}$ feet.............1 pace.

MARINER'S MEASURE.

Table used by mariners in calculating distances on water, and the speed of vessels.

9 in................1 span. | 7½ cables......1 mile or knot.
8 spans, or 6 ft....1 fathom. | 51 ft. nearly....1 "   "   "
120 fath.......1 cable's length. | 3 miles.........1 league.

NOTE.—The number of knots of the log line run off in half a minute indicates the number of knots of distance a vessel goes per hour.

WEIGHTS AND MEASURES.

SURVEYOR'S LONG MEASURE,

For measuring boundaries of land, areas, railroads, canals.

7 92/100 Inches.........1 link. | 4 Rods..............1 chain.
25 Links................1 rod. | 80 Chains...........1 mile.

EQUIVALENTS.

mi.	ch.	rd.	l.	in.
1 =	80 =	320 =	8000 =	63360
	1 =	4 =	100 =	792
		1 =	25 =	198
			1 =	7.92

SCALE OF UNITS:—7.92, 25, 4, 80.

10 chains long by 1 broad, or 10 square chains, 1 acre.

GUNTER'S CHAIN, which is the unit of measure used by surveyors, is 66 feet long, consisting of 100 links.

Measurements are recorded in chains and hundredths. Latterly a steel measuring tape 100 feet long, with each foot divided into tenths, is used by engineers as a substitute for the cumbersome chain.

NOTE.—By scientific persons and revenue officers, the inch is divided into *tenths, hundredths, etc.* Among mechanics, the inch is divided into *eighths.* The division of the inch into 12 parts, called lines, is not now in use.

A standard English mile, which is the measure that we use, is 5280 feet in length, 1760 yards, or 320 rods. A strip, one rod wide and one mile long, is two acres. By this it is easy to calculate the quantity of land taken up by roads, and also how much is wasted by fences.

TABLE

For Geographical and Astronomical Calculations.

1 Geographic mile..........................1.15 statute miles
3   "   "   ....1 league.
60   "   "   or 69.16 statute miles.1 degree.
360 Degrees..................................Circumference of the earth.

NOTE.—The earth's circumference is 24,855½ miles, nearly. The nautical mile is 6075¾ feet, or 795¾ feet longer than the common mile.

SURFACE OR SQUARE MEASURE

Used in ascertaining the extent of surfaces, such as land, boards, plastering, paving, etc.

TABLE.

144 Square Inches (*sq. in.*)...1 square foot, *sq. ft.*
9 Square Feet....................1 square yard, *sq. yd.*
30¼ Square Yards................1 sq. rod or perch, *sq. rd.; P.*
160 Square Rods.................1 acre, *A.*
640 Acres.......................1 square mile, *sq. mi.*

sq. mi.	*A.*	*sq. rd.*	*sq. yd.*	*sq. ft.*	*sq. in.*
1 =	640 =	102400 =	3097600 =	27878400 =	4014489600
	1 =	160 =	4840 =	43560 =	6272640
		1 =	30¼ =	272¼ =	39204
			1 =	9 =	1296
				1 =	144

SCALE OF UNITS:—144, 9, 30¼, 40, 4, 640.

Measure 209 feet on each side, and you have a square acre within an inch.

NOTE.—The following gives the comparative size, in square yards, of acres in different countries:

English acre, 4840 square yards; Scotch, 6150; Irish, 7840; Hamburgh, 11,545; Amsterdam, 9722; Dantzic, 6650; France (hectare), 11,960; Prussia (morgen), 3053.

This difference should be borne in mind in reading of the products per acre in different countries. Our land measure is that of England.

ARTIFICERS estimate their work as follows:
By the *square foot;* as in glazing, stone-cutting, etc.
By the *square yard,* or by the *square* of 100 square feet; as in plastering, flooring, roofing, paving, etc.
One thousand shingles, averaging 4 inches wide, and laid 5 inches to the weather, are estimated to be a *square.*

SURVEYORS' SQUARE MEASURE,

Used for measuring the area or contents of fields, farms, and government lands.

TABLE.

625 Square Links (*sq. l.*)............1 pole, *P.*
16 Poles.................................1 square chain, *sq. ch.*
10 Square Chains..................1 acre, *A.*
640 Acres..............................1 square mile, *sq. mi.*
36 Square Miles (6 miles sq.)....1 township, *Tp.*

EQUIVALENTS.

Tp.	*sq. mi.*	*A.*	*sq. ch.*	*P.*	*sq. l.*
1 =	36 =	23040 =	230400 =	3686400 =	2304000000
	1 =	640 =	6400 =	102400 =	64000000
		1 =	10 =	160 =	10000
			1 =	16 =	1000
				1 =	625

SCALE OF UNITS:—625, 16, 10, 640, 36.

The acre is the unit of land measure.

GOVERNMENT LAND MEASURE.

A township—36 sections, each a mile square.
A section—640 acres.
A quarter section, half a mile square—160 acres.
An eighth section, half a mile long, north and south, and a quarter of a mile wide—80 acres.
A sixteenth section, a quarter of a mile square—40 acres.

The sections are all numbered 1 to 36, commencing at the northeast corner, thus:

6	5	4	3	2	NW|NE SW|SE
7	8	9	10	11	12
18	17	16*	15	14	13
19	20	21	22	23	24
30	29	28	27	26	25
31	32	33	34	35	36

The sections are all divided into quarters, which are named by the cardinal points, as in section 1. The quarters are divided in the same way. The description of a forty-acre lot would read: The south half of the west half of the south-west quarter of section 1 in township 24, north of range 7 west, or as the case might be; and sometimes will fall short, and sometimes overrun the number of acres it is supposed to contain.

CONTENTS OF FIELDS AND LOTS.

For the convenience of farmers and others who desire lay off small lots of land for sale, or to ascertain the amount of land in fields, the following table is prepared, and will be found accurate:

$52\tfrac{1}{8}$ ft. square or $2722\tfrac{1}{2}$ square ft. $= \tfrac{1}{16}$ of an acre.
$73\tfrac{3}{4}$ " " " 5445 " " $= \tfrac{1}{8}$ " " "
$104\tfrac{1}{3}$ " " " 10890 " " $= \tfrac{1}{4}$ " " "
$120\tfrac{3}{4}$ " " " 14520 " " $= \tfrac{1}{3}$ " " "
$147\tfrac{7}{12}$ " " " 21780 " " $= \tfrac{1}{2}$ " " "
$208\tfrac{3}{4}$ " " " 43560 " " $= 1$ "

10 rods × 16 rods = 1 A	110 feet × 396 feet = 1 A
8 " × 20 " = 1 "	60 " × 726 " = 1 "
5 " × 32 " = 1 "	120 " × 363 " = 1 "
4 " × 40 " = 1 "	240 " × $181\tfrac{1}{2}$ " = 1 "
5 yds. × 968 yds. = 1 "	200 " × $108\tfrac{9}{10}$ " = $\tfrac{1}{2}$ "
10 " × 484 " = 1 "	100 " × $145\tfrac{2}{10}$ " = $\tfrac{1}{3}$ "
20 " × 242 " = 1 "	100 " × $108\tfrac{9}{10}$ " = $\tfrac{1}{4}$ "
40 " × 121 " = 1 "	
80 " × $60\tfrac{1}{2}$ " = 1 "	25 ft. × 100 ft. = .0574 "
70 " × $69\tfrac{1}{7}$ " = 1 "	25 " × 110 " = .0631 "
	25 " × 120 " = .0688 "
220 feet × 198 feet = 1 "	25 " × 125 " = .0717 "
440 " × 99 " = 1 "	25 " × 150 " = .109 "

TABLE

Showing the number of Rails, Stakes and Posts required for 10 rods of Post-and-Rail Fence.

Length of rail. ft.	Length of panel. ft.	No. of panels.	No. of posts.	Number of rails.	
				6 rails high.	7 rails high.
10	8	$20\tfrac{5}{8}$	21	124	145
12	10	$16\tfrac{1}{2}$	17	99	116
14	12	$13\tfrac{3}{4}$	14	83	96
$16\tfrac{1}{2}$	$14\tfrac{1}{2}$	$11\tfrac{1}{3}$	12	68	79

NOTE.—In arranging the above table 12 inches lap have been allowed. The greater the lap, the stronger and more durable the fence. To ascertain the number of rails, etc., for any desired length of fence, multiply the numbers given in the above table by the length in feet, and point off one figure from the left, and you have the desired result.

TROY WEIGHT

Is used for weighing gold, silver, platina, and precious stones, except diamonds; also in philosophical experiments.

TABLE.

24 grains (*gr.*) make 1 pennyweight, *pwt.*
20 pwt. make 1 ounce, *oz.*
12 oz.   "   1 pound, *lb.*

EQUIVALENTS.

lb.		oz.		pwt.		gr.
1	=	12	=	240	=	5760
		1	=	20	=	480
				1	=	24

Scale of units:—24, 20, 12.

NOTE.—Troy Weight contains 5760 grains to the pound, or 1240 grains less than the avoirdupois pound. In mixing medicines, not liquid, apothecaries use Troy grains.

DIAMOND WEIGHT TABLE.

16 parts........................1 grain.
4 gr ....1 carat.
1 K.............................3¼ Troy grains, nearly 3½.

The word carat is used to express the fineness of gold, and means $\frac{1}{24}$ part. Pure gold is said to be 24 carats fine; if there be 22 parts of pure gold and 2 parts of alloy, it is said to be 22 carats fine. The standard of American coin is nine-tenths pure gold, and is worth $20.67. What is called the *new standard*, used for watch cases, etc., is 18 carats fine. The term carat is also applied to a weight of 3¼ grains Troy, used in weighing diamonds; it is divided into 4 parts, called *grains;* 4 grains Troy are thus equal to 5 grains diamond weight.

PAPER AND BOOKS.

The following denominations of measure are used by the paper manufacturer, book and stationery trade.

TABLE.
24 Sheets..........1 quire.
20 Quires..........1 ream.
2 Reams..........1 bundle.
5 Bundles.......1 bale.
1 Bale contains 200 quires or 4800 sheets.

SIZES OF PAPER.

Paper manufactured to order can be made any desired size. The following are regular or trade sizes, and can generally be found in stock at any of the wholesale paper houses.

WRITING PAPERS—FLAT CAP.

Name.	Size, In.	Name.	Size, In.
Law Blank	13×16	Medium	18 ×23
Flat Cap	14×17	Royal	19 ×24
Crown	15×19	Super Royal	20 ×28
Demy	16×21	Imperial	22 ×30
Folio Post	17×22	Elephant	$22\frac{1}{4}$ ×$27\frac{3}{4}$
Check Folio	17×24	Columbia	23 ×$33\frac{1}{4}$
Double Cap	17×28	Atlas	26 ×33
Ex. Size Folio	19×23	Doub. Elephant	26 ×40

WRITING PAPERS—FOLDED.

Name.	Size, In.	Name.	Size, In.
Billet Note	6 × 8	Letter	10 × 16
Octavo Note	7 × 9	Commerc'l Let.	11 × 17
Commercial Note.	8 × 10	Packet Post	11½ × 18
Packet Note	9 × 11	Ex. Pack. Post.	11½ × 18½
Bath Note	8½ × 14	Foolscap	12½ × 16

PRINTING PAPER.

Used in Printing Newspapers and Books.

Name.	Size, In.	Name.	Size, In.
Medium	19 × 24	Double Medium	24 × 38
Royal	20 × 25	Double Royal	26 × 40
Super Royal	22 × 28	Doub. Sup. Royal	28 × 42
Imperial	22 × 32	" " "	29 × 43
Medium-and-half	24 × 30	Broad Twelves	23 × 41
Small Doub. Med.	24 × 36	Double Imperial	32 × 46

COPYING.

In the copying of papers, manuscripts, and documents for official record, clerks and copyists are usually paid by the folio.

A folio varies in quantity in different States and sections of the world, but is generally estimated from 75 to 100 words.

PRINTING—TYPE-SETTING.

Printers generally charge for setting type, or the composition of matter, as it is technically termed, by the number of *ems* it contains, rated by the 1000 *ems;* an *em* is the square of the body of the type.

PRINTING—PRESS-WORK.

Press-work is generally charged by the token of 250 impressions, or 125 sheets, printed on both sides. The value or cost of press-work depends upon the style, quantity, and quality of ink used.

MISCELLANEOUS TABLES.
FOR COUNTING CERTAIN ARTICLES.

12 Units or pieces..........1 dozen.
20 " " ....1 score.
12 Dozen..................1 gross.
12 Gross................ .1 great-gross.

BOOKS.

Names and Sizes as classified by Publishers.

The number of folds and pages in a single sheet when manufactured.

Name of book.	When a sheet is folded into leaves.	Contain.
Folio.....................	2 leaves....	4 pages.
Quarto or 4to...............	4 "	....8 "
Octavo or 8vo..............	8 "	....16 "
Duodecimo or 12 mo........	12 "	....24 "
16 mo.....................	32 "	....64 "
18 " *...................	18 "	....36 "
24 "	24 "	....48 "
32 "	32 "	....64 "

* NOTE.—This book is an 18mo., there being 36 pages to the sheet. The terms *folio, quarto, octavo, etc.*, denote the number of leaves in which a sheet of paper is folded. The marks A, B, C; 1, 2, 3; 1A, 2A; 1*, 2*, etc., occasionally found at the bottom of pages, are what printers term *signature marks*, thus, 3*, being printed for the direction of binders in folding the sheets.

Printers and stationers generally procure their supplies of paper by the quantity, the cost per ream varying according to the quality and weight.

The table on page 264 will be found invaluable in making up their estimate for small job work, as it shows at a glance the *cost per quire* of paper purchased at 15 to 30 cents per pound, and weighing from 10 to 60 pounds to the ream.

EXAMPLE.—What does *one quire* of paper cost purchased at 23 cents per pound, and weighing 40 pounds to the ream? *Ans.* 46 cents.

EXPLANATION.—Refer to the weight column on the left and the purchasing price at the top, and you have the cost per quire shown in the purchasing column on the line with the weight.

SHOEMAKER'S MEASURE.

3 Barleycorns or sizes..................1 inch.

Number one, *children's measure*, is $4\frac{3}{8}$ inches, and that every additional number calls for an increase of $\frac{1}{3}$ of an inch in length. Number one, *adults' measure*, is $8\frac{1}{2}$ inches long, with a gradual increase of $\frac{1}{3}$ of an inch for additional numbers, so that, for example, number ten measures $11\frac{1}{2}$ inches. This measure corresponds to the number of the *last*, and not to the length of the *sole*.

TABLE
To Ascertain the Cost of One Quire of Paper.

Wg't of Ream. lbs.	\multicolumn{16}{c}{Purchasing Price in Cents per Pound.}
	15	16	17	18	19	20	21	22	23	24	25	26	27	28	29	30
10	07.5	08	08.5	09	09.5	10	10.5	11	11.5	12	12.5	13	13.5	14	14.5	15
12	09	09.6	10.2	10.8	11.4	12	12.6	13.2	13.8	14.4	15	15.6	16.2	16.8	17.4	18
14	10.5	11.2	11.9	12.6	13.3	14	14.7	15.4	16.1	16.8	17.5	18.2	18.9	19.6	20.3	21
16	12	12.8	13.6	14.4	15.2	16	16.8	17.6	18.4	19.2	20	20.8	21.6	22.4	23.2	24
18	13.5	14.4	15.3	16.2	17.1	18	18.9	19.8	20.7	21.6	22.5	23.4	24.3	25.2	26.1	27
20	15	16	17	18	19	20	21	22	23	24	25	26	27	28	29	30
22	16.5	17.6	18.7	19.8	20.9	22	23.1	24.2	25.3	26.4	27.5	28.6	29.7	30.8	31.9	33
24	18	19.2	20.4	21.6	22.8	24	25.2	26.4	27.6	28.8	30	31.2	32.4	33.6	34.8	36
26	19.5	20.8	22.1	23.4	24.7	26	27.3	28.6	29.9	31.2	32.5	33.8	35.1	36.4	37.7	39
28	21	22.4	23.8	25.2	26.6	28	29.4	30.8	32.2	33.6	35	36.4	37.8	39.2	40.6	42
30	22.5	24	25.5	27	28.5	30	31.5	33	34.5	36	37.5	39	40.5	42	43.5	45
32	24	25.6	27.2	28.8	30.4	32	33.6	35.2	36.8	38.4	40	41.6	43.2	44.8	46.4	48
34	25.5	27.2	28.9	30.6	32.3	34	35.7	37.4	39.1	40.8	42.5	44.2	45.9	47.6	49.3	51
36	27	28.8	30.6	32.4	34.2	36	37.8	39.6	41.4	43.2	45	46.8	48.6	50.4	52.2	54
38	28.5	30.4	32.3	34.2	36.1	38	39.9	41.8	43.7	45.6	47.5	49.4	51.3	53.2	55.1	57
40	30	32	34	36	38	40	42	44	46	48	50	52	54	56	58	60
42	31.5	33.6	35.7	37.8	39.9	42	44.1	46.2	48.3	50.4	52.5	54.6	56.7	58.8	60.9	63
44	33	35.2	37.4	38.6	41.8	44	46.2	48.4	50.6	52.8	55	57.2	59.4	61.6	63.8	66
46	34.5	36.8	39.1	41.4	43.7	46	48.3	50.6	52.9	55.2	57.5	59.8	62.1	64.4	66.7	69
48	36	38.4	40.8	43.2	45.6	48	50.4	52.8	55.2	57.6	60	62.4	64.8	67.2	69.6	72
50	37.5	40	42.5	45	47.5	50	52.5	55	57.5	60	62.5	65	67.5	70	72.5	75
52	39	41.6	44.2	46.8	49.4	52	54.6	57.2	59.8	62.4	65	67.6	70.2	72.8	75.4	78
54	40.5	43.2	45.9	48.6	51.3	54	56.7	59.4	62.1	64.8	67.5	70.2	72.9	75.6	78.3	81
56	42	44.8	47.6	50.4	53.2	56	58.8	61.6	64.4	67.2	70	72.8	75.6	78.4	81.2	84
58	43.5	46.4	49.3	52.2	56.1	58	60.9	63.8	66.7	69.6	72.5	75.4	78.3	81.2	84.1	87
60	45	48	51	54	57	60	63	66	69	72	75	78	81	84	87	90

MEASUREMENT OF TIME.
TIME IS THE MEASURE OF DURATION

We have in this engraving a representation of the magnificent transit instrument used in the Paris Observatory. It is made on the same model as the celebrated one in the Observatory at Greenwich. These, and the one at the National Observatory at Washington, are the finest in the world The instrument is used for the purpose of determining the instant of time a heavenly body passes, : makes a transit across the meridian.

MEASUREMENT OF TIME.
TABLE.

60 seconds	1 minute.
60 minutes	1 hour.
24 hours	1 day.
7 days	1 week.
28 days	1 lunar month.
28, 29, 30, or 31 days	1 cal. month.
12 cal. months	1 year.
365 days*	1 com. year.
366 days*	1 leap year.*
365¼ days	1 Julian year.
365 d., 5 h., 48 m., 49 s.	1 solar year.
365 d., 6 h., 9 m., 12 s.	1 siderial year.
10 years	1 decade.
10 decades, or 100 years	1 century.

EQUIVALENTS.

yr.		da.		hr.		min.		sec.
1	=	365¼	=	8766	=	525960	=	31557600
		1	=	24	=	1440	=	86400
				1	=	60	=	3600
						1	=	60

Scale of units:—60, 60, 24, 365¼.

* NOTE.—The common year thus consists of 365 days. Once in 4 years, however, one day is added to February, making 366 days; and thus, each year averages 365¼ days. The longest year is called Bissextile, or Leap year. Centuries divisible by 400, and other years divisible by 4, are leap years.

In business transactions 30 days are considered 1 month. The civil day begins and ends at 12 o'clock, midnight.

In dating events, astronomers calculate the day as beginning and ending at 12 o'clock, noon.

CIRCULAR MEASURE,
Or, Divisions of the Circle,

Used in Astronomy, Geography, Navigation, and Surveying; also for calculating the differences of time.

TABLE.

60 seconds (")	1 minute....'
60 '	1 degree.....°
30°	1 sign........S.
12 S., or 360°	1 circle......C.*

EQUIVALENTS.

C.		S.		°		'		"
1	=	12	=	360	=	21600	=	1296000
		1	=	30	=	1800	=	108000
				1	=	60	=	3600
						1	=	60

Scale of units:—60, 60, 30, 12.

*A semi-circumference is ½ of a circle, 180°
A quadrant " ¼ " " " 90°
A sextant " ⅙ " " " 60°

The greatest distance across a circle is called its diameter. The distance around it is called its *circumference*. Any part of the circumference is called an *arc*.

LONGITUDE AND TIME.
TABLE.

For a difference of	There is a difference of
1° in Long............................	4 m. in Time.
1' "	4 sec. "
1" "	1/15 sec. "
1 hr. in Time........................	15° in Long.
1 m. "	15' "
1 sec. "	15"

NOTE.—*Add* difference of time for places *east* and subtract it for places *west* of any given place.

HOW TO ASCERTAIN
The Difference of Time between Cities.

BASIS OF CALCULATION.

360 degrees = 1 revolution of the earth, or 1 day.
1440 minutes = 1   "   "   "   "   " 1   "

1440 ÷ 360 = 4 minutes, or 1 degree.

Refer to your map and notice the difference in degrees of longitude between places. Multiply the number of degrees by 4; the product will be the difference in time.

Degrees of Longitude East, time increases.
"   "   "   West, " decreases.

PROBLEM.

When it is 12 o'clock noon at Washington, what time is it at Boston? Ans., 12.24.

Per map, difference in degrees, 6 east, which increase the time 6 × 4 = 24 differences in time, or 24 minutes past 12.

PROBLEM.

When it is 12 o'clock noon at Washington, what is the time in San Francisco, California? Ans., 8.58 A. M.

Per map, difference 45½ degrees, West; 45½ × 4 = 182 min., or 3h. 2m. less. Ans., 8 o'clock, 58 min., A. M.

A telegram sent from Washington to San Francisco at 12 M. will be received at 9 o'clock A. M. (i. e., *three hours before it is sent*), calculating San Francisco time.

TABLE

For ascertaining the number of days between two dates.

Jan.	Feb.	Mar.	Apl.	May	June	July	Aug.	Sept.	Oct.	Nov.	Dec.
1	32	60	91	121	152	182	213	244	274	305	335
2	33	61	92	122	153	183	214	245	275	306	336
3	34	62	93	123	154	184	215	246	276	307	337
4	35	63	94	124	155	185	216	247	277	308	338
5	36	64	95	125	156	186	217	248	278	309	339
6	37	65	96	126	157	187	218	249	279	310	340
7	38	66	97	127	158	188	219	250	280	311	341
8	39	67	98	128	159	189	220	251	281	312	342
9	40	68	99	129	160	190	221	252	282	313	343
10	41	69	100	130	161	191	222	253	283	314	344
11	42	70	101	131	162	192	223	254	284	315	345
12	43	71	102	132	163	193	224	255	285	316	346
13	44	72	103	133	164	194	225	256	286	317	347
14	45	73	104	134	165	195	226	257	287	318	348
15	46	74	105	135	166	196	227	258	288	319	349
16	47	75	106	136	167	197	228	259	289	320	350
17	48	76	107	137	168	198	229	260	290	321	351
18	49	77	108	138	169	199	230	261	291	322	352
19	50	78	109	139	170	200	231	262	292	323	353
20	51	79	110	140	171	201	232	263	293	324	354
21	52	80	111	141	172	202	233	264	294	325	355
22	53	81	112	142	173	203	234	265	295	326	356
23	54	82	113	143	174	204	235	266	296	327	357
24	55	83	114	144	175	205	236	267	297	328	358
25	56	84	115	145	176	206	237	268	298	329	359
26	57	85	116	146	177	207	238	269	299	330	360
27	58	86	117	147	178	208	239	270	300	331	361
28	59	87	118	148	179	209	240	271	301	332	362
29		88	119	149	180	210	241	272	302	333	363
30		89	120	150	181	211	242	273	303	334	364
31		90		151		212	243		304		365

NOTE.—To find from the above table the number of days between two dates, we give the following—

RULE I.—*When the dates are in the same year, subtract the number of days of the earlier date from the number of days of the later date; the result will be the number of days required.*

II. *When the dates are in consecutive years, subtract the number of days of the earlier date from 365, and add to the remainder the number of days of the later date; the result will be the number of days required.*

When the year is a leap year, add one day to the result.

TABLE

Showing the number of days from any day in one month to the same day in any other.

From \ To	Jan	Feb	Mar.	April.	May.	June.	July.	Aug.	Sept.	Oct.	Nov.	Dec.
Jan.........	365	31	59	90	120	151	181	212	243	273	304	334
Feb.........	334	365	28	59	89	120	150	181	212	242	273	303
March.......	306	337	365	31	61	92	122	153	184	214	245	275
April........	275	306	334	365	30	61	91	122	153	183	214	244
May..........	245	276	304	335	365	31	61	92	123	153	184	214
June.........	214	245	273	304	334	365	30	61	92	122	153	183
July.........	184	215	243	274	304	335	365	31	62	92	123	153
Aug..........	153	184	212	243	273	304	334	365	31	61	92	122
Sept.........	122	153	181	212	242	273	303	334	365	30	61	91
Oct..........	92	123	151	182	212	243	273	304	335	365	31	61
Nov..........	61	92	120	151	181	212	242	273	304	334	365	30
Dec..........	31	62	90	121	151	182	212	243	274	304	335	365

NOTE.—Find in the left-hand column the month from any day of which you wish to compute the number of days to the same day in any other month; then follow the line along until under the desired month, and you have the required number of days.

EXAMPLE.—How many days from March 15 to July 15?
Ans., 122 days.

In leap-year, when the month of February occurs in the calculation, one day extra must be added.

EXAMPLE.—1876. How many days from January 13 to May 13? Ans., Per table, 120; one day added for leap-year = 121 days.

ASTRONOMICAL CALCULATIONS

A scientific method of telling immediately what day of the week any date transpired or will transpire, from the commencement of the Christian Era, for the term of three thousand years.

MONTHLY TABLE.

The ratio to add for each month will be found in the following table:

Ratio of June is..............0	Ratio of October is..........3
Ratio of September is......1	Ratio of May is................4
Ratio of December is.......1	Ratio of August is............5
Ratio of April is................2	Ratio of March is.............6
Ratio of July is.................2	Ratio of February is........6
Ratio of January is..........3	Ratio of November is......6

NOTE.—On Leap Year the ratio of January is 2, and the ratio of February is 5. The ratio of the other ten months do not change on Leap Years.

CENTENNIAL TABLE.

The ratio to add for each century will be found in the following table:

Christian Era.						
200,	900,	1800,	2200,	2600,	3000,	ratio is...........0
300,	1000,	......	......	......	......	ratio is...........6
400,	1100,	1900,	2300,	2700,	......	ratio is...........5
500	1200,	1600	2000,	2400,	2800,	ratio is......... 4
600	1300,	......	......	......	......	ratio is......... 3
000,	700,	1400,	1700,	2100,	2500, 2900,	ratio is..........2
100,	800,	1500,	......	......	......	ratio is..........1

ASTRONOMICAL CALCULATIONS. 271

NOTE.—The figure opposite each century is its ratio; thus the ratio for 200, 900, etc., is 0. To find the ratio of any century, first find the century in the above table, then run the eye along the line until you arrive at the end; the small figure at the end is its ratio.

METHOD OF OPERATION.

RULE.*—*To the given year add its fourth part, rejecting the fractions; to this sum add the day of the month; then add the ratio of the month and the ratio of the century. Divide this sum by 7; the remainder is the day of the week, counting Sunday as the first, Monday as the second, Tuesday as the third, Wednesday as the fourth, Thursday as the fifth, Friday as the sixth, Saturday as the seventh; the remainder for Saturday will be 0 or zero.*

EXAMPLE 1.—Required the day of the week for the 4th of July, 1810.

To the given year, which is...................10
Add its fourth part, rejecting fractions............... 2
Now add the day of the month, which is............ 4
Now add the ratio of July, which is... 2
Now add the ratio of 1800, which is.... 0

Divide the whole sum by 7.   7 | 18—4
                                   2

We have 4 for a remainder, which signifies the fourth day of the week, or Wednesday.

---

*When dividing the year by 4, always leave off the centuries. We divide by 4 to find the number of Leap Years.

Note.—In finding the day of the week for the present century, no attention need be paid to the *centennial ratio,* as it is 0.

Example 2.—Required the day of the week for the 2d of June, 1805.

To the given year, which is............................... 5
Add its fourth part, rejecting fractions............... 1
Now add the day of the month, which is 2
Now add the ratio of June, which is..................... 0

Divide the whole sum by 7.   7 | 8—1
                                                     1

We have 1 for a remainder, which signifies the first day of the week, or Sunday.

The Declaration of American Independence was signed July 4, 1776. Required the day of the week.

To the given year, which is...............................76
Add its fourth part, rejecting fractions...............19
Now add the day of the month, which is 4
Now add the ratio of July, which is..................... 2
Now add the ratio of 1700, which is................... 2

Divide the whole sum by 7.   7 | 103—5
                                                 14

We have 5 for a remainder, which signifies the fifth day of the week, or Thursday.

The Pilgrim Fathers landed on Plymouth Rock Dec. 20, 1620. Required the day of the week.

To the given year, which is. 20
Add its fourth part, rejecting fractions............... 5
Now add the day of the month, which is............ 20
Now add the ratio of December which is............ 1
Now add the ratio of 1600, which is................... 4

Divide the whole sum by 7.   7 | 50--1
                                7

We have 1 for a remainder, which signifies the first day of the week, or Sunday.

On what day will happen the 8th of January, 1815? *Ans.* Sunday.

On what day will happen the 4th of May, 1810?

On what day will happen the 3d of December, 1423? *Ans.* Friday.

On what day of the week were you born?

The earth revolves round the sun once in 365 days, 5 hours, 48 minutes, 48 seconds; this period is, therefore, a *Solar* year. In order to keep pace with the solar year, in our reckoning, we make every fourth to contain 366 days, and call it Leap Year. Still greater accuracy requires, however, that the leap day be dispensed with three times in every 400 years. Hence, every year (except the centennial years) that is divisible by 4 is a *Leap Year*, and every centennial year that is divisible by 400 is also a *Leap Year*. The next centennial year that will be a Leap Year is 2000

MONEY OF THE UNITED STATES

Is the measure of value of all kinds, such as property, merchandise, services, etc. It is the medium of exchange in business.

COIN or SPECIE is metal stamped and authorized by government to be used as money.

PAPER MONEY consists of notes issued by the United States Treasury and banks, and used as money. United States money is the legal currency of the United States.

U. S. MONEY.

10 mills (M.)............1 cent........*ct.* or ¢.
10¢...................1 dime........
10 dimes...............1 dollar......D. or $.
10 dollars..............1 eagle.......E.

NOTE.—The *mill* is not coined.

The COIN of the United States consists of *gold, silver, nickel* and *bronze*, and as fixed by the "New Coinage Act" of 1873 is as follows:

GOLD. The double-eagle, eagle, half-eagle, quarter-eagle, three-dollar, and one-dollar pieces.

SILVER. The *Trade*-dollar, half-dollar, quarter-dollar, the twenty-cent, and the ten-cent pieces.

NICKEL. The five-cent and three-cent pieces.

BRONZE. The one-cent piece.

NOTE.—The term shilling is frequently used in the United States in stating the price of articles, and it indicates old divisions or equivalents of parts of the dollar. Its value varies in different States as follows: In the New England States, and in Indiana, Illinois, Missouri, Mississippi, Texas, Virginia, Kentucky, and Tennessee, 1s.=16⅔ cts., and $1. =6s.; in New York, Ohio, Michigan, and North Carolina, 1s.=12½ cts. and $1.=8s.; in Pennsylvania, New Jersey, Delaware, and Maryland, 1s.=13⅓ cts., and $1.=7½s.; in Georgia and South Carolina, 1s.= 21¾ cts., and $1.=4⅗s. These rates are liable to variations by custom; as, in Illinois, the shilling is rated frequently at 12½.

CANADA MONEY consists, like United States money, of dollars and cents. The Canada coins are twenty-cent, ten-cent, five-cent, *silver;* and one-cent, *bronze.*

MONEY OF FRANCE.

The money of account of France is the Franc of 100 Centimes or 1000 Sous, and is arranged on the decimal system.

The principal gold coins in circulation are as follows: Louis d'or, Forty Franc piece, Twenty Franc piece, and Six Franc piece.

The principal silver coins are, the Crown, ½ Crown, ¼ Crown, Five Franc piece, Two Franc piece, Franc, ½ Franc of 50 Centimes, ¼ Franc of 25 Centimes.

The par value of the Franc is 19 cents and 3 mills, but its commercial value varies according to the fluctuations of the money market. When exchange is quoted at say 5.17½, it is understood to mean that 5 Francs and 17½ Centimes are equal to the gold dollar. The gold cost of any given number of francs is, therefore, ascertained by dividing that number by the *quotation* or rate of exchange. For example, the cost of 9463 Francs at 5.21½ = 9463 fr. ÷ 5.215 fr. = $1814.57. The premium on gold is then added to find the cost in currency.

MONEY OF THE GERMAN EMPIRE.

The German Empire has, within the past few years, issued a new coin called the Mark, which is now adopted as the money of account of that nation.

The gold and silver coins are quite numerous, embracing, as they do, those in use in about twenty-two States, and are as follows:

Gold.—Ducat, Quintuple Ducat, Five Thaler piece, Ten Thaler piece, Double d'or, ¼ Caroline, ½ Caroline, Caroline, Five Gilder piece, Twenty Mark piece, Ten Mark piece, and Twelve Mark piece.

Silver.—Mark, Thaler, Double Thaler, Crown Thaler, ⅛ Thaler, Double Gilder, Florin, ½ Florin, Twelve Grote piece, Grote, Rix Dollar, Crown, Kreutzer Groschen, 6 Pfen, 1 Schilling, 48 Schilling piece, 30 Kreutzer piece, 8 Schilling piece.

In commercial transactions the Mark is not generally reckoned at its par value, but is governed by the quotations which range at present between 91 cents and $1.00 for 4 Marks. The market value of the Mark, in gold, is found by dividing the *quotation* by 4. For example, 400 Marks at $.95½ = $.955 ÷ 4 m. ×400 = $95.50. The premium on gold is then added to find the cost in currency.

ARBITRATION OF EXCHANGE.

The method of finding the value, in gold and currency, of the moneys of the principal nations has already been briefly explained; but the fluctuations of foreign exchange sometimes render it to the advantage of the merchant to remit *indirectly*. For example, suppose it is required to pay a debt due in France, and that the balance of trade between Great Britain and the United States is in favor of the latter nation. This, of course, would cause a decline in the value of the £ sterling in United States money. Now, if there is no corresponding difference in value between the moneys of Great Britain and France, it would be cheaper first to purchase sterling and then remit through London. Such exchanges are termed *direct* and *indirect*, as the case will indicate, and are treated under the rule of

ARBITRATION OF EXCHANGE.

The limits of a work of this kind prevent an extended elucidation of this subject, and it is not thought best to present any set rules for memorizing.

In the following examples, with solutions, *indirect exchange* will be found much simplified. Indeed, when taken in the order of the different nations involved, there will appear but little distinction from the method of working ordinary exchange.

EXAMPLE 1.—When sterling is quoted at $4.83½ U. S. money, and 25.73 fr., French, how many francs are equal to the gold dollar by indirect exchange?
Solution.—If £1 = $4.83½, and also 25.73 fr., $4.83½ must equal 25.73 fr. Therefore 25.73 fr. ÷ $4.835 = 5.32 fr., ans.

EXAMPLE 2.—When French exchange is quoted 5.16½ fr., and sterling 25.73 fr., what is the value in gold of the £ sterling by indirect exchange?
Solution.—Since $1=5.16½ fr., and £1=25.73 fr., therefore 25.73 fr.÷ 5.165 = $4.98.

EXAMPLE 3.—What would be the gain in sterling on $6000 by remitting through France, Germany, and Netherlands, with the following quotations: $1= 5.18 fr.; 1.22 fr. = 1 mark; 1.71 marks = 1 guil.; 11.8 guil.=£1, when the £ sterling by direct exchange is quoted $4.85?
Solution.—As France is the first country, $6000 × 5.18 fr. = 31080 fr. ÷1.22 fr. = 24575.475 m. ÷ 1.71 m. = 14847.94 guil. ÷ 11.8 guil. = £1262.537=£1262 7s. 4¾+d., indirect exchange. $6000÷$4.85=£1237 1134 = £1237 2s. 3¼+d. direct exchange. £1262 7s. 4¾d. less £1237 2s. 3¼d. = £25 5s. 1½d. gain.

Table exhibiting the Values, in United States Money, of the Pure Gold or Silver representing respectively the Monetary Units and Standard Coins of Foreign Countries, in compliance with the Act of March 3, 1873.

The values given in this Table were estimated by the Director of the Mint, and published January 1, 1888.

Country.	Monetary Unit.	Standard.	Value in U.S. Money.
Argentine Republic	Peso fuerte	Gold	$0.96,5
Austria	Florin	Silver	.34,5
Belgium	Franc	Gold and silver	.19,3
Bolivia	Boliviano	Single	.69,9
Brazil	Milreis of 1000 reis	Gold	.54,6
British America	Dollar	Gold	1.00
Bogota	Peso	Gold	.91,2
Central America	Dollar	Silver	.91,8
Chili	Peso	Gold	.91,2
Cuba	Peso	Gold	.92,6
Denmark	Crown	Gold	.26,8
Ecuador	Sucre	Silver	.69,9
Egypt	Pound of 100 piasters	Gold	4.94,3
France	Franc	Gold and silver	.19,3
Great Britain	Pound sterling	Gold	4.86,8½

VALUE OF FOREIGN COINS.

Greece	Drachma	Gold and silver	.19,3
German Empire	Mark	Gold	.23,8
Hayti	Gourde	Silver	.96,5
Japan	Yen	Gold	.99,7
India	Rupee of 16 annas	Silver	.38,2
Italy	Lira	Gold and silver	.19,3
Liberia	Dollar	Gold	1.00
Mexico	Dollar	Silver	.75,9
Netherlands	Florin	Silver	.40,2
Norway	Crown	Gold	.26,8
Paraguay	Peso	Gold	1.00
Peru	Dollar	Silver	.69,9
Porto Rico	Peso	Gold	.92,5
Portugal	Milreis of 1000 reis	Gold	1.08
Russia	Rouble of 100 copecks	Silver	.55,9
Sandwich Islands	Dollar	Gold	1.00
Spain	Peseta of 100 centimes	Gold and silver	.19,3
Sweden	Crown	Gold	.26,8
Switzerland	Franc	Gold and silver	.19,3
Tripoli	Mahbub of 20 piasters	Silver	.63
Tunis	Piaster of 16 caroube	Silver	.04,4
Turkey	Piaster	Gold	.69,9
U. S. of Colombia	Peso	Silver	.69,9
Venezuela	Bolivar	Silver	.14

GOLD AND CURRENCY.

Gold is usually represented as rising and falling, but being the standard of value, it does not vary. The variation is in the currency substituted for gold or specie; hence, when gold is said to be at a premium, the currency or circulating medium is made the standard, while it is in fact below par.

TABLE
Showing the Comparative Value of Gold and Currency.

When $1 in Gold is sold for Currency at				The Discount on Currency is		The Amount in Gold which can be bought for $100 in Currency.	
1.01	or	1 per cent.	Prem.	1.00	per cent.	$99.99 or	99 $\frac{1}{100}$
1.05	"	5	"	4.77	"	95.23 "	95 $\frac{5}{21}$
1.10	"	10	"	9.10	"	90.90 "	90 $\frac{10}{11}$
1.15	"	15	"	13.04	"	86.96 "	86 $\frac{22}{23}$
1.20	"	20	"	16.67	"	83.33 "	83 $\frac{1}{3}$
1.25	"	25	"	20.00	"	80.00 "	80
1.30	"	30	"	23.08	"	76.92 "	76 $\frac{12}{13}$
1.33⅓	"	33⅓	"	25.00	"	75.00 "	75
1.40	"	40	"	28.58	"	71.42 "	71 $\frac{3}{7}$
1.50	"	50	"	33.33	"	66.66 "	66⅔
1.60	"	60	"	37.50	"	62.50 "	62½
1.66⅔	"	66⅔	"	40.00	"	60.00 "	60
1.70	"	70	"	41.18	"	58.82 "	58 $\frac{14}{17}$
1.80	"	80	"	44.45	"	55.55 "	55 $\frac{5}{9}$
1.90	"	90	"	47.37	"	52.63 "	52 $\frac{12}{19}$
2.00	"	100	"	50.00	"	50.00 "	50
2.50	"	150	"	60.00	"	40.00 "	40
5.00	"	400	"	80.00	"	20.00 "	20
7.50	"	650	"	86.67	"	13.33 "	13½
10.00	"	900	"	90.00	"	10.00 "	10
50.00	"	4900	"	98.00	"	2.00 "	2
100.00	"	9900	"	99.00	"	1.00 "	1

TO ASCERTAIN HOW MUCH GOLD
Can be Bought for a Stated Amount of Currency.

RULE.—*Add two ciphers to the amount of currency (in dollars), and divide by 100, increased by the premium rate on gold; the quotient will be the gold sum.*

BANK ACCOUNTS.

HOW TO TRANSACT
BUSINESS WITH BANKS.

BANK ACCOUNTS.
HINTS TO MANY AND PRACTICAL ADVICE TO THOUSANDS.

It is the belief of many observant philosophic persons who are well on their way through life, that if people generally knew more they would behave better, though few, if any of them, believe that knowledge and morality are synonymous terms. Acting on this conviction, the following "Hints to those who keep Bank Accounts" have been suggested by a gentleman well qualified by general intelligence and long practical experience to advise the young and untaught of the several matters.

HINTS TO THOSE WHO KEEP BANK ACCOUNTS.

1. If you wish to open an account with a bank, pro-

vide yourself with a proper introduction. Well-managed banks do not open accounts with strangers.

2. Do not draw a check unless you have the money in bank or in your possession to deposit. Don't test the courage or generosity of your bank by presenting, or allowing to be presented, your check for a larger sum than your balance.

3. Do not draw a check and send it to a person out of the city, expecting to make it good before it can possibly get back. Sometimes telegraphic advice is asked about such checks.

4. Do not exchange checks with *anybody*. This is soon discovered by your bank; it does your friend no good and discredits you.

5. Do not give your check to a friend with the condition that he is not to use it until a certain time. He is sure to betray you, for obvious reasons. Do not take an out-of-town check from a neighbor, pass it through your bank without charge, and give him your check for it. You are sure to get caught.

6. Do not give your check to a stranger. This is an open door for fraud, and if your bank loses through you, it will not feel kindly to you.

7. When you send your checks out of the city to pay bills, write the name and residence of your payee, thus: Pay to Jno. Smith & Co., *of Boston*. This will put your bank on its guard, if presented at the counter.

8. Don't commit the folly of supposing that, because you trust the bank with your money, the bank ought to trust you by paying your overdrafts.

9. Don't suppose you can behave badly in one bank and stand well with the others. You forget there is a Clearing House.

10. Don't quarrel with your bank. If you are not treated well, go somewhere else, but don't go and leave your discount line unprotected. Don't think it unreasonable if your bank declines to discount an accommodation note. Have a clear definition of an accommodation note—in the meaning of a bank, it is a note for which no value has passed from the endorser to the drawer.

11. If you want an accommodation note discounted, tell your bank frankly that it is not, in their definition, a business note. If you take a note from a debtor with an agreement, verbal or written, that it is to be renewed in whole or in part, and if you get that note discounted, and then ask to have a new one discounted to take up the old one, tell your bank all about it.

12. Don't commit the folly of saying that you will guarantee the payment of a note which you have *already endorsed.*

13. Give your bank credit for being intelligent generally and understanding its own business particularly. It is much better informed, probably, than you suppose.

14. Don't try to convince your bank that the paper or security which has already been declined is better than the bank supposes. This is only chaff.

15. Don't quarrel with a teller because he does not pay you in money exactly as you wish. As a rule, he does the best he can.

16. In all your intercourse with bank officers, treat them with the same courtesy and candor that you would expect and desire if the situations were reversed.

17. Don't send ignorant and stupid messengers to bank to transact your business.

INTEREST—COMMERCIAL RULE.

Commercial Year.—Calculations based on 360 days to the year, or 30 days to the month.

SIX PER CENT.

Rule.—*Multiply the given number of dollars by the number of days of interest required; divide the product by 6, and point off three figures from the right.*

Note—If cents appear in the principal, it will be necessary to point off five figures.

The result is $\frac{1}{73}$ more than the true interest based on the calculation of 365 days per annum.

To ascertain the true amount, it will only be necessary to deduct $\frac{1}{73}$ from the result obtained.

Example.—What is the interest on $1000 for 219 days?

Process.— 1000 × 219 = 219000
219000 ÷ 6 = 36500

Point off three figures from the right, gives $36.50 interest.

To ascertain the true interest (365 days), subtract $\frac{1}{73}$, thus; 36.50 ÷ 73 = 50. $36.50 — 50 = $36.

Having ascertained the interest at 6 per cent., that for 7, 8, and 9 per cent. is readily found, by adding to it $\frac{1}{6}$, $\frac{1}{3}$, $\frac{1}{2}$, etc., etc.

To find the interest at any per cent., divide the

interest procured at 6 per cent. by 6, and multiply the amount by the required rate.

VALUABLE INTEREST RULES.

Basis Commercial Year of 360 days, or 30 days per month.

4 *per cent.*—Multiply the principal by the required number of days, divide by 9, and point off.

5 *per cent.*—Multiply by the number of days, and divide by 72.

6 *per cent.*—Multiply by the number of days, divide by 6, and point off three figures from the right.

8 *per cent.*—Multiply by the number of days, and divide by 45.

9 *per cent.*—Multiply by the number of days, divide by 4, and point off three figures from the right.

10 *per cent.*—Multiply by the number of days, and divide by 36.

12 *per cent.*—Multiply by the number of days, divide by 3, and point off three figures from the right.

15 *per cent.*—Multiply by the number of days, and divide by 24.

18 *per cent.*—Multiply by the number of days, divide by 2, and point off three figures from the right.

20 *per cent.*—Multiply by the number of days, and divide by 18.

☞ The interest in each case will be in dollars and cents.

INTEREST TABLES.

Rate, 10 *per cent. per annum of* 360 *days.*

These Tables are arranged with a view to supply a want long existing among the great majority of business men and accountants, and are specially designed to supersede the numerous high-priced Interest Tables now before the public.

CALCULATIONS.

The basis is at the decimal rate of 10 per cent. per annum on the commercial year of 360 days, and the

TIME

for which our calculations of interest are made is from 1 to 30 days, and from 1 to 12 months, on amounts ranging from $10 to $10,000, thereby meeting the wants of the capitalist as well as those of the more moderate tradesman.

THE CALCULATIONS

are of the most minute accuracy, being carried out in each instance to the fraction of *one tenth of one mill;*

hence, in summing up the amount of interest, it will always be necessary to point off *four figures* from the left.

When amounts from $1 to $10 appear in the principal, the interest is obtained by taking the sum in the interest column opposite the required amounts from $10 to $90, and point off five figures from the left.

CENTS IN THE PRINCIPAL.

When there are cents in the principal in excess of 50, add $1.00, if less, reject them; in the calculation of interest this is in accordance with usual custom. It is also customary when the fraction of interest is 5 mills, or in excess, to add one cent; when less, drop it.

TO ASCERTAIN INTEREST

on the basis of 365 *days per annum.*

Subtract $\frac{1}{73}$ from the results obtained by the calculations on the basis of 360 days per annum, which is equivalent to a reduction of $1\frac{1}{3}$ cents for every dollar of interest.

TO ASCERTAIN INTEREST

at any rate other than 10 *per cent.*

Multiply the amount of interest obtained from the Tables by the required rate, and point off five figures from the left.

INSTRUCTION.

It will only be necessary to trace the work of the following examples to enable any one to become expert in the

USE OF THE TABLES.

In calculating interest, refer to the Table containing at its head the number of days or months for which interest is required, and opposite the principal in the column of dollars will be found the interest in dollars, cents, mills and tenths of mills.

Example I.—Ascertain the interest on $3570 for 28 days at 10 per cent.

See Table, page 292.

Dollar Column.	Interest.
$3000	$23.3333
500	3.8889
70	.5444
$3570	$27.7666

Interest at 10 per cent., 28 days.

To find the Interest on above amount at **6 per cent**:

Interest at 10 per cent.,	$27.7666
Multiply by required rate,	6 per cent.
	$16.65996

Point off five figures from the left and you will have the interest at 6 per cent., $16.66.

OR, THUS:

By removing the decimal point one place to the left we have the interest at 1 per cent.; hence, by simply multiplying by the required rate, the product will be the desired interest.

INTEREST TABLES. 289

Example II.—Ascertain the interest on $2007 for 7 months, 17 days, at 10 per cent.
See Table, page 293.

	Dollar Column.		Interest.
	$2000		$116.6667
	7	($70.—$4.0833)	.4083

See Table, page 291, 17 days.

	Dollar Column.		
	$2000		9.4444
	7	($70.— .3306)	.0330
	$2007		$126.5524

Interest 7 months, 17 days, 10 per cent.

To find the interest on above amount at 7 per cent:

Interest at 10 per cent., $126.5524
Multiply by required rate, 7 per cent.
$88.58668

Point off five figures from the left, and we have the interest at 7 per cent., $88.59.

Orton & Sadler's Interest Tables.

Year of 360 days. Rate, 10% per annum.

Dolls.	1 day.	2 days.	3 days.	4 days.	5 days.	6 days.	7 days.	8 days.	9 days.	10 days.
10000	$2.77.78	$5.55.56	$8.33.33	$11.11.11	$13.88.89	$16.66.67	$19.44.44	$22.22.22	$25.00.00	$27.77.78
9000	2.50.00	5.00.00	7.50.00	10.00.00	12.50.00	15.00.00	17.50.00	20.00.00	22.50.00	25.00.00
8000	2.22.22	4.44.44	6.66.67	8.88.89	11.11.11	13.33.33	15.55.56	17.77.78	20.00.00	22.22.22
7000	1.94.44	3.88.89	5.83.33	7.77.78	9.72.22	11.66.67	13.61.11	15.55.56	17.50.00	19.44.44
6000	1.66.67	3.33.33	5.00.00	6.66.67	8.33.33	10.00.00	11.66.67	13.33.33	15.00.00	16.66.67
5000	1.38.89	2.77.78	4.16.67	5.55.56	6.94.44	8.33.33	9.72.22	11.11.11	12.50.00	13.88.89
4000	1.11.11	2.22.22	3.33.33	4.44.44	5.56.56	6.66.67	7.77.78	8.88.89	10.00.00	11.11.11
3000	.83.33	1.66.67	2.50.00	3.33.33	4.16.67	5.00.00	5.83.33	6.66.67	7.50.00	8.33.33
2000	.55.56	1.11.11	1.66.67	2.22.22	2.77.78	3.33.33	3.88.89	4.44.44	6.00.00	5.55.56
1000	.27.78	.55.56	.83.33	1.11.11	1.38.89	1.66.67	1.94.44	2.22.22	2.50.00	2.77.78
900	.25.00	.50.00	.75.00	1.00.00	1.25.00	1.50.00	1.75.00	2.00.00	2.25.00	2.50.00
800	.22.22	.44.44	.66.67	.88.89	1.11.11	1.33.33	1.55.56	1.77.78	2.00.00	2.22.22
700	.19.44	.38.89	.58.33	.77.78	.97.22	1.16.67	1.36.11	1.55.56	1.75.00	1.94.44
600	.16.67	.33.33	.50.00	.66.67	.83.33	1.00.00	1.16.67	1.33.33	1.50.00	1.66.67
500	.13.89	.27.78	.41.67	.55.56	.69.44	.83.33	.97.22	1.11.11	1.25.00	1.38.89
400	.11.11	.22.22	.33.33	.44.44	.55.56	.66.67	.77.78	.88.89	1.00.00	1.11.11
300	.08.33	.16.67	.25.00	.33.33	.41.67	.50.00	.58.33	.66.67	.75.00	.83.33
200	.05.50	.11.11	.16.67	.22.22	.27.78	.33.33	.38.89	.44.44	.50.00	.55.56
100	.02.78	.05.56	.08.33	.11.11	.13.89	.16.67	.19.44	.22.22	.25.00	.27.78
90	.02.50	.05.00	.07.50	.10.00	.12.50	.15.00	.17.50	.20.00	.22.50	.25.00
80	.02.22	.04.44	.06.67	.08.89	.11.11	.13.33	.15.56	.17.78	.20.00	.22.22
70	.01.94	.03.89	.05.83	.07.78	.09.72	.11.67	.13.61	.15.56	.17.50	.19.44
60	.01.67	.03.33	.05.00	.06.67	.08.33	.10.00	.11.67	.13.33	.15.00	.16.67
50	.01.39	.02.78	.04.17	.05.56	.06.94	.08.33	.09.72	.11.11	.12.50	.13.89
40	.01.11	.02.22	.03.33	.04.44	.05.56	.06.67	.07.78	.08.89	.10.00	.11.11
30	.00.83	.01.67	.02.50	.03.33	.04.17	.05.00	.05.83	.06.67	.07.50	.08.33
20	.00.56	.01.11	.01.67	.02.22	.02.78	.03.33	.03.89	.04.44	.05.00	.05.56
10	.00.28	.00.56	.00.83	.01.11	.01.39	.01.67	.01.94	.02.22	.02.50	.02.78

Orton & Sadler's Interest Tables.
Year of 360 days. Rate, 10% per annum.

Dolls.	11 days.	12 days.	13 days.	14 days.	15 days.	16 days.	17 days.	18 days.	19 days.	20 days.
10000	$30.55 56	$33.33 33	$36.11 11	$38.88 89	$41.66 67	$44.44 44	$47.22 22	$50.00 00	$52.77 77	$55.55 56
9000	27.50 00	30.00 00	32.50 00	35.00 00	37.50 00	40.00 00	42.50 00	45.00 00	47.50 00	50.00 00
8000	24.44 44	26.66 67	28.88 89	31.11 11	33.33 33	35.55 56	37.77 78	40.00 00	42.22 22	44.44 44
7000	21.38 89	23.33 33	25.27 78	27.22 22	29.16 67	31.11 11	33.05 56	35.00 00	36.94 44	38.88 88
6000	18.33 33	20.00 00	21.66 67	23.33 33	25.00 00	26.66 67	28.33 33	30.00 00	31.66 67	33.33 31
5000	15.27 78	16.66 67	18.05 56	19.44 44	20.83 33	22.22 22	23.61 11	25.00 00	26.38 89	27.77 78
4000	12.22 22	13.33 33	14.44 44	15.55 56	16.66 67	17.77 78	18.88 89	20.00 00	21.11 11	22.22 22
3000	9.16 67	10.00 00	10.83 33	11.66 67	12.50 00	13.33 33	14.16 67	15.00 00	15.83 33	16.66 66
2000	6.11 11	6.66 67	7.22 22	7.77 78	8.33 33	8.88 89	9.44 44	10.00 00	10.55 56	11.11 11
1000	3.05 56	3.33 33	3.61 11	3.88 89	4.16 67	4.44 44	4.72 22	5.00 00	5.27 78	5.55 56
900	2.75 00	3.00 00	3.25 00	3.50 00	3.75 00	4.00 00	4.25 00	4.50 00	4.75 00	5.00 00
800	2.44 44	2.66 67	2.88 89	3.11 11	3.33 33	3.55 56	3.77 78	4.00 00	4.22 22	4.44 44
700	2.13 89	2.33 33	2.52 78	2.72 22	2.91 67	3.11 11	3.30 56	3.50 00	3.69 44	3.88 89
600	1.83 33	2.00 00	2.16 67	2.33 33	2.50 00	2.66 67	2.83 33	3.00 00	3.16 67	3.33 33
500	1.52 78	1.66 67	1.80 56	1.94 44	2.08 33	2.22 22	2.36 11	2.50 00	2.63 89	2.77 73
400	1.22 22	1.33 33	1.44 44	1.55 56	1.66 67	1.77 78	1.88 99	2.00 00	2.11 11	2.22 22
300	.91 67	1.00 00	1.08 33	1.16 67	1.25 00	1.33 33	1.41 67	1.50 00	1.58 33	1.66 67
200	.61 11	.66 67	.72 22	.77 78	.83 33	.88 89	.94 44	1.00 00	1.05 56	1.11 11
100	.30 56	.33 33	.36 11	.38 89	.41 67	.44 44	.47 22	.50 00	.52 78	.55 56
90	.27 50	.30 00	.32 50	.35 00	.37 50	.40 00	.42 50	.45 00	.47 50	.50 00
80	.24 44	.26 67	.28 89	.31 11	.33 33	.35 56	.37 78	.40 00	.42 22	.44 44
70	.21 39	.23 33	.25 28	.27 22	.29 17	.31 11	.33 06	.35 00	.36 94	.38 89
60	.18 33	.20 00	.21 67	.23 33	.25 00	.26 67	.28 33	.30 00	.31 67	.33 33
50	.15 28	.16 67	.18 06	.19 44	.20 83	.22 22	.23 61	.25 00	.26 39	.27 78
40	.12 22	.13 33	.14 44	.15 56	.16 67	.17 78	.18 89	.20 00	.21 11	.22 22
30	.09 17	.10 00	.10 83	.11 67	.12 50	.13 33	.14 17	.15 00	.15 83	.16 67
20	.06 11	.06 67	.07 22	.07 78	.08 33	.08 89	.09 44	.10 00	.10 56	.11 11
10	.03 06	.03 33	.03 61	.03 89	.04 17	.04 44	.04 72	.05 00	.05 28	.05 56

Urton & Sadler's Interest Tables.
Year of 360 days.   Rate, 10% per annum.

Dolls.	21 days.	22 days.	23 days.	24 days.	25 days.	26 days.	27 days.	28 days.	29 days.	30 days.
10000	$58.33 33	$61.11 11	$63.88 89	$66.66 66	$69.44 44	$72.22 22	$75.00 00	$77.77 78	$80.55 55	$83.33 33
9000	52.50 00	55.00 00	57.50 00	60.00 00	62.50 00	65.00 00	67.50 00	70.00 00	72.50 00	75.00 00
8000	46.66 67	48.88 89	51.11 11	53.33 34	55.55 55	57.77 78	60.00 00	62.22 22	64.44 44	66.66 66
7000	40.83 33	42.77 78	44.72 22	46.66 67	48.61 11	50.55 56	52.50 00	54.44 44	56.38 89	58.33 33
6000	35.00 00	36.66 67	38.33 33	40.00 00	41.66 67	43.33 33	45.00 00	46.66 67	48.33 33	50.00 00
5000	29.16 67	30.55 56	31.94 44	33.33 33	34.72 22	36.11 11	37.50 00	38.88 89	40.27 78	41.66 67
4000	23.33 33	24.44 44	25.55 66	26.66 67	27.77 78	28.88 89	30.00 00	31.11 11	32.22 22	33.33 33
3000	17.50 00	18.33 33	19.16 67	20.00 00	20.83 33	21.66 67	22.50 00	23.33 33	24.16 67	25.00 00
2000	11.66 67	12.22 22	12.77 78	13.33 33	13.88 89	14.44 44	15.00 00	15.55 56	16.11 11	16.66 67
1000	5.83 33	6.11 11	6.38 89	6.66 67	6.94 44	7.22 22	7.50 00	7.77 78	8.05 56	8.33 33
900	5.25 00	5.50 00	5.75 00	6.00 00	6.25 00	6.50 00	6.75 00	7.00 00	7.25 00	7.50 00
800	4.66 67	4.88 89	5.11 11	5.33 33	5.55 56	5.77 78	6.00 00	6.22 22	6.44 44	6.66 67
700	4.08 33	4.27 78	4.47 22	4.66 67	4.86 11	5.05 56	5.25 00	5.44 44	5.63 89	5.83 33
600	3.50 00	3.66 67	3.83 33	4.00 00	4.16 67	4.33 33	4.50 00	4.66 67	4.83 33	5.00 00
500	2.91 67	3.05 56	3.19 44	3.33 33	3.47 22	3.61 11	3.75 00	3.88 89	4.02 78	4.16 67
400	2.33 33	2.44 44	2.55 56	2.66 67	2.77 78	2.88 89	3.00 00	3.11 11	3.22 22	3.33 33
300	1.75 00	1.83 33	1.91 67	2.00 00	2.08 33	2.16 67	2.25 00	2.33 33	2.41 67	2.50 00
200	1.16 67	1.22 22	1.27 78	1.33 33	1.38 89	1.44 44	1.50 00	1.55 56	1.61 11	1.66 67
100	.58 33	.61 11	.63 89	.66 67	.69 44	.72 22	.75 00	.77 78	.80 56	.83 33
90	.52 50	.55 00	.57 50	.60 00	.62 50	.65 00	.67 50	.70 00	.72 50	.75 00
80	.46 67	.48 89	.51 11	.53 33	.55 56	.57 78	.60 00	.62 22	.64 44	.66 67
70	.40 83	.42 78	.44 72	.46 67	.48 61	.50 56	.52 50	.54 44	.56 39	.58 33
60	.35 00	.36 67	.38 33	.40 00	.41 67	.43 33	.45 00	.46 67	.48 33	.50 00
50	.29 17	.30 56	.31 94	.33 33	.34 72	.36 11	.37 50	.38 89	.40 28	.41 67
40	.23 33	.24 44	.25 56	.26 67	.27 78	.28 89	.30 00	.31 11	.32 22	.33 33
30	.17 50	.18 33	.19 17	.20 00	.20 83	.21 67	.22 50	.23 33	.24 17	.25 00
20	.11 67	.12 22	.12 78	.13 33	.13 89	.14 44	.15 00	.15 66	.16 11	.16 67
10	.05 83	.06 11	.06 39	.06 67	.06 94	.07 22	.07 50	.07 78	.08 06	.08 33

Orton & Sadler's Interest Tables.
Year of 360 days.   Rate, 10% per annum.

Dolls.	1 month, or 30 days.	2 months, or 60 days.	3 months, or 90 days.	4 months, or 120 days.	5 months, or 150 days.	6 months, or 180 days.	7 months, or 210 days.	8 months, or 240 days.	Dolls.
10000	$83.33 33	$166.66 67	$250.00 00	$333.33 33	$416.66 67	$500.00 00	$583.33 33	$666.66 67	10000
9000	75.00 00	150.00 00	225.00 00	300.00 00	375.00 00	450.00 00	525.00 00	600.00 00	9000
8000	66.66 67	133.33 33	200.00 00	266.66 67	333.33 33	400.00 00	466.66 67	533.33 33	8000
7000	58.33 33	116.66 67	175.00 00	233.33 33	291.66 67	350.00 00	408.33 33	466.66 67	7000
6000	50.00 00	100.00 00	150.00 00	200.00 00	250.00 00	300.00 00	350.00 00	400.00 00	6000
5000	41.66 67	83.33 33	125.00 00	166.66 67	208.33 33	250.00 00	291.66 67	333.33 33	5000
4000	33.33 33	66.66 67	100.00 00	133.33 33	166.66 67	200.00 00	233.33 33	266.66 67	4000
3000	25.00 00	50.00 00	75.00 00	100.00 00	125.00 00	150.00 00	175.00 00	200.00 00	3000
2000	16.66 67	33.33 33	50.00 00	66.66 67	83.33 33	100.00 00	116.66 67	133.33 33	2000
1000	8.33 33	16.66 67	25.00 00	33.33 35	41.66 67	50.00 00	58.33 33	66.66 67	1000
900	7.50 00	15.00 00	22.50 00	30.00 00	37.50 00	45.00 00	52.50 00	60.00 00	900
800	6.66 67	13.33 33	20.00 00	26.66 67	33.33 33	40.00 00	46.66 67	53.33 33	800
700	5.83 33	11.66 67	17.50 00	23.33 33	29.16 67	35.00 00	40.83 33	46.66 67	700
600	5.00 00	10.00 00	15.00 00	20.00 00	25.00 00	30.00 00	35.00 00	40.00 00	600
500	4.16 67	8.33 33	12.50 00	16.66 67	20.83 33	25.00 00	29.16 67	33.33 33	500
400	3.33 33	6.66 67	10.00 00	13.33 33	16.66 67	20.00 00	23.33 33	26.66 67	400
300	2.50 00	5.00 00	7.50 00	10.00 00	12.50 00	15.00 00	17.50 00	20.00 00	300
200	1.66 67	3.33 33	5.00 00	6.66 67	8.33 33	10.00 00	11.66 67	13.33 33	200
100	.83 33	1.66 67	2.50 00	3.33 33	4.16 67	5.00 00	5.83 33	6.66 67	100
90	.75 00	1.50 00	2.25 00	3.00 00	3.75 00	4.50 00	5.25 00	6.00 00	90
80	.66 67	1.33 33	2.00 00	2.66 67	3.33 33	4.00 00	4.66 67	5.33 33	80
70	.58 33	1.16 67	1.75 00	2.33 33	2.91 67	3.50 00	4.08 33	4.66 67	70
60	.50 00	1.00 00	1.50 00	2.00 00	2.50 00	3.00 00	3.50 00	4.00 00	60
50	.41 67	.83 33	1.25 00	1.66 67	2.08 33	2.50 00	2.91 67	3.33 33	50
40	.33 33	.66 67	1.00 00	1.33 33	1.66 67	2.00 00	2.33 33	2.66 67	40
30	.25 00	.50 00	.75 00	1.00 00	1.25 00	1.50 00	1.75 00	2.00 00	30
20	.16 67	.33 33	.50 00	.66 67	.83 33	1.00 01	1.16 67	1.33 33	20
10	.08 33	.16 67	.25 00	.33 33	.41 67	.50 00	.52 33	.64 67	10

Orton & Sadler's Interest Tables.

Rate 10 per cent., 360 days per annum.

Dolls.	9 months, or 270 days.	10 months, or 300 days.	11 months, or 330 days.	12 months, or 360 days.	Dolls.
10000	$750.00|00	$833.33|33	$916.66|67	$1000.00|00	10000
9000	675.00|00	750.00|00	825.00|00	900.00|00	9000
8000	600.00|00	666.66|67	733.33|33	800.00|00	8000
7000	525.00|00	583.33|33	641.66|67	700.00|00	7000
6000	450.00|00	500.00|00	550.00|00	600.00|00	6000
5000	375.00|00	416.66|67	458.33|33	500.00|00	5000
4000	300.00|00	333.33|33	366.66|67	400.00|00	4000
3000	225.00|00	250.00|00	275.00|00	300.00|00	3000
2000	150.00|00	166.66|67	183.33|33	200.00|00	2000
1000	75.00|00	83.33|33	91.66|67	100.00|00	1000
900	67.50|00	75.00|00	82.50|00	90.00|00	900
800	60.00|00	66.66|67	73.33|33	80.00|00	800
700	52.50|00	58.33|33	64.16|67	70.00|00	700
600	45.00|00	50 00|00	55.00|00	60.00|00	600
500	37.50|00	41.66|67	45.83|33	50.00|00	500
400	30.00|00	33.33|33	36.66|67	40.00|00	400
300	22.50|00	25.00|00	27.50|00	30.00|00	300
200	15.00|00	16.66|67	18.33|33	20.00|00	200
100	7.50|00	8.33|33	9.16|67	10.00|00	100
90	6 75|00	7.50|00	8.25|00	9.00|00	90
80	6.00|00	6.66|67	7.33|33	8.00|00	80
70	5.25|00	5.83|33	6.41|67	7.00|00	70
60	4.50|00	5.00|00	5.50|00	6.00|00	60
50	3.75|00	4.16|67	4.58|33	5.00|00	50
40	3.00|00	3.33|33	3.66|67	4.00|00	40
30	2.25|00	2.50|00	2.75|00	3.00|00	30
20	1.50|00	1.66|67	1.83|33	2.00|00	20
10	.75|00	.83|33	.91|67	1.00|00	10

COMPOUND INTEREST TABLE,

Showing the amount of $1.00 *at Compound Interest, from* 1 *to* 20 *years. Rate* 5 *to* 10 *per cent.*

Years.	5 Per Cent.	6 Per Cent.	7 Per Cent.	10 Per Cent.
1	1.050000	1.060000	1.070000	1.100000
2	1.102500	1.123600	1.144900	1.210000
3	1.157625	1.191016	1.225043	1.331000
4	1.215506	1.262477	1.310796	1.464100
5	1.276282	1.338226	1.402552	1.610510
6	1.340096	1.418519	1.500730	1.771561
7	1.407100	1.503630	1.605781	1.948717
8	1.477455	1.593848	1.718186	2.143589
9	1.551328	1.689479	1.838459	2.357948
10	1.628895	1.790848	1.967151	2.593742
11	1.710339	1.898299	2.104852	2.853117
12	1.795856	2.012196	2.252192	3.138428
13	1.885649	2.132928	2.409845	3.452271
14	1.979932	2.260904	2.578534	3.797498
15	2.078928	2.396558	2.759031	4.177248
16	2.182875	2.540352	2.952164	4.594973
17	2.292018	2.692773	3.158815	5.054470
18	2.406619	2.854339	3.379932	5.559917
19	2.526950	3.025599	3.616527	6.115909
20	2.653298	3.207135	3.869684	6.727500

N. B. In the calculations of Compound Interest, much labor will be saved by use of the above Table.

RULE.—*Refer to the Table, ascertain the amount of* $1.00 *for the given time at the specified rate, and multiply same by the principal.*

Example.—What will be the amount of $600 for 15 years, at 6 per cent. Compound Interest?

Process.—See Table. Amount $1. for 15 years at 6 per cent $2.396558
   Amount........................ 600
Am't of $600 for 15 yrs. at 6 per cent..$1437.934800

TO FIND THE COMPOUND INTEREST,
Time and Rate being given.

RULE.—*Subtract the principal invested from the amount.*

Example.—What is the Compound Interest on $600 for 15 years, at 6 per cent.?

From the above Example, $600, 15 years, 6 per cent., Comp. Int., we have the amount.....$1437.93
   Principal invested...................... 600.
   Compound Interest $837.93

Process.—$600 invested at Compound Interest, for 15 years, at 6 per cent., will amount to $1437.93.

TABLE
Showing in how many YEARS *a given principal will* DOUBLE ITSELF.

RATE.	AT SIMPLE INTEREST.	AT COMPOUND INTEREST.	
		Compounded Yearly.	Compounded Half-Yearly.
4½	22.22	15.748	15.576
5	20.00	14.207	14.036
6	16.67	11.896	11.725
7	14.29	10.245	10.075
8	12.50	9.006	8.837
9	11.11	8.043	7.874
10	10.00	7.273	7.+

TABLE OF INTEREST RATES FOR THE U. S.
Penalties for Usury, and Statute Limitations.

STATES AND TERRITORIES.	Legal rate of interest. Per Ct.	Rate allowed by contract. Per Ct.	PENALTIES FOR USURY.	Statute Limitations. Open Accts. Yrs.	Notes. Yrs.
Alabama	8	8	Forfeiture of entire interest.	3	6
Arizona	10	Any			
Arkansas	6	10		2,3	7&21
California	7	Any		2	4
Colorado	10	"		2	4
Connecticut	6	6	Forfeiture of entire interest.	6	17
Dakota	7	12		6	15
Delaware	6	6	Forfeiture of the principal...	3	6&20
District of Columbia.	6	10	Forfeiture of entire interest.	3	3
Florida	8	Any		5	5
Georgia	7	8	Forfeiture of excess	3	3
Idaho	10	18			
Illinois	6	8	Forfeiture of entire interest.	5	6
Indiana	6	8	Forfeiture of excess	6	20
Iowa	6	10	Forfeiture of entire interest; 10 per ct. of it to sch. fund.	5	10
Kansas	7	12	Forfeiture of entire interest.	3	5
Kentucky	6	6	Forfeiture of excess	1, 2	7
Louisiana	5	8	Forfeiture of entire interest.	3	5
Maine	6	Any		6	6
Maryland	6	6	Forfeiture of excess	3	3
Massachusetts	6	Any		6&20	6&20
Michigan	7	10	Forfeiture of excess	6	6
Minnesota	7	10	" " "	6	6
Mississippi	6	10	" " "	3	6
Missouri	6	10	Forfeiture of entire interest.	5	10
Montana	10	Any			
Nebraska	7	10	Forfeiture of entire interest.	4	5
Nevada	10	Any			
New Hampshire	6	6	Forfeiture of three times the excess and costs	6	6
New Jersey	6	6	Forfeiture of entire interest.	6	16
New Mexico	6	Any			
New York	6	6	Forfeiture of excess	6	6
North Carolina	6	8	Forfeiture of entire interest.	3	3
Ohio	6	8	Forfeiture of excess	6	15
Oregon	8	10		6	6
Pennsylvania	6	6		6	6, 20
Rhode Island	6	Any		6(?)	6

TABLE OF INTEREST RATES—*Continued.*

STATES AND TERRITORIES.	Legal rate of interest.	Rate allowed by contract.	PENALTIES FOR USURY.	Statute Limitations	
				Open Accts.	Notes.
	Per Ct.	Per Ct.		Yrs.	Yrs.
South Carolina	7	7		6	6
Tennessee	6	6	Forfeiture of excess; fine and imprisonment	6	6
Texas	8	12		2	4
Utah	10	Any			
Vermont	6	6	Forfeiture of excess	6	6, 14
Virginia	6	6	Forfeiture of excess, in action of *equity*	5	5, 20
Washington	10	Any			
West Virginia	6	8	Forfeiture of excess	5	5, 20
Wisconsin	7	10	Forfeiture of entire interest.	10	6
Wyoming	12	Any			

NOTES.—The legal rate of interest in Canada, Nova Scotia, and Ireland, is 6 per cent.; England and France, 5 per cent.

When the rate is not specified, the legal rate is always understood and so allowed by the courts.

Debts of all kinds draw interest from the time they become due, but not before, unless specified.

METRIC SYSTEM

[*From Sadler's Counting House Arithmetic,*]

THE *Metric System* is a system used in measuring *Length, Surface, Volume, Capacity,* and *Weight.* Its basis is called the *Meter,* which is equal to *one ten-millionth* ($\frac{1}{10000000}$) part of a quadrant of the earth's circumference, or of the distance from either Pole to the Equator, as determined by the measurement of an Arc of the Meridian.

The Metric System is used in France, Germany, Spain, Portugal, Belgium, Holland, Italy, Austria, Switzerland, Sweden, Norway, Denmark, Greece, Mexico, Brazil, Turkey, Guatemala, Venezuela, Ecuador, Chili, San Salvador, Argentine Republic, and the United States of Colombia. In some of those countries it is used exclusively, while in others its use is legalized. In 1864 its use was allowed in Great Britain, if agreed to by the parties concerned. In 1866, by virtue of an Act of Congress, its use was authorized in the United States. The five-cent piece now coined is one-fiftieth of a Meter in diameter, and 5 grains in weight.

The *Higher Denominations* are termed as follows:

Deka.	*Hekto.*	*Kilo.*	*Myria.*
10	100	1000	10000.

The above terms are Greek Numerals, and represent the higher Integers or Multiples.

The *Lower Denominations* are termed as follows:

Deci.	*Centi.*	*Milli.*
.1 or $\frac{1}{10}$	.01 or $\frac{1}{100}$	.001 or $\frac{1}{1000}$

The above terms are Latin Ordinals, and represent the Decimals or Submultiples. The value of a denomination is expressed by prefixing a Greek Numeral or a Latin Ordinal to the Unit involved. Thus, Dekameter means 10 Meters, and Hektometer means 100 Meters, etc. So also, Decimeter means 1-10th of a Meter, and Centimeter means 1-100th of a Meter, etc.

The *Unit of Length* is called the *Meter* (meeter). Its extent is equal to 39.37079 United States inches, or about 328 feet. The Standard Meter is marked by two parallel lines on a bar of platinum, which is deposited in the Archives of Paris. The Equivalents of the Multiplier and Sub-Multiplier of Units of Lengths are shown in the following

TABLE OF LINEAR MEASURE.

1 Millimeter (mm.)		= .03937079 in.
10 Millimeters	= 1 Centimeter (cm.)	= .3937079 in.
10 Centimeters	= 1 Decimeter (dm.)	= 3.937079 in.
10 Decimeters	= 1 Meter (m.)	= 3.2808992 ft.
10 Meters	= 1 Dekameter (dm.)	= 32.808992 ft.
10 Dekameters	= 1 Hektometer (hm.)	= 19.927817 rd.
10 Hektometers	= 1 Kilometer (km.)	= .6213824 ml.
10 Kilometers	= 1 Myriameter (mm.)	= 6.213824 ml.

The *Unit of Area* is the *Are* (air). It is a square Dekameter, and equals 119.6 sq. yd., or .0247 A. The equivalents are shown in the following

TABLE OF SQUARE MEASURE.

1 Centiare	= 1550.0591 sq. in.
100 Centiare = 1 *Are*	= 119.6034 sq. yds.
100 Ares = 1 Hektare	= 2.47114 acres.

Surfaces other than land are measured by the Square Meter and Square Decimeter. The denominations are shown in the following

TABLE.

100 Square Millimeters	= 1 Square Centimeter.
100 Square Centimeters	= 1 Square Decimeter.
100 Square Decimeters	= 1 Square Meter.

METRIC SYSTEM. 301

The ***Unit of Volume*** is the ***Stere***.

It is used principally in the measurement of wood. The other denominations are shown in the following

TABLE OF SOLID MEASURE.

1 Decistere		=	6102.705151 cubic inches.
10 Decisteres	= 1 *Stere*	=	35.31658 cubic feet.
10 Steres	= 1 Deikastere	=	353.1658 cubic feet.

Solids other than wood are generally measured by the ***Cubic Meter*** or multiplier thereof. The denominations are shown in the following

TABLE.

1000 Cubic Centimeters = 1 Cubic Decimeter (cu. dm.)
1000 Cubic Millimeters = 1 Cubic Centimeter (cu. cm.)
1000 Cubic Decimeters = 1 Cubic *Meter* (cu. m.)

The ***Unit of Capacity*** is the ***Liter***. It is a Cubic Decimeter, or a Cube, whose equal dimensions are .1 Meter × .1 Meter × .1 Meter. The denominations are shown in the following Table:

MEASURES OF CAPACITY.

1 Milliliter (ml.)		=	.0610270515	cu. in.
10 Milliliters	= 1 Centiliter (cl.)	=	.610270515	"
10 Centiliters	= 1 Deciliter (dl.)	=	6.10270515	"
10 Deciliters	= 1 Liter (L.)	=	61.0270515	"
10 Liters	= 1 Dekaliter (D.L.)	=	610.270515	"
10 Dekaliters	= 1 Hektoliter (H.L.)	=	6102.70515	"
10 Hektoliters	= 1 Kiloliter (K.L.)	=	61027.0515	"
10 Kiloliters	= 1 Myrialiter (M.L.)	=	610270.515	"

The *Liter* is the standard unit of both Liquid and Dry Measures. It is equal to 1.05673 quarts Liquid Measure, and .9081 of a quart Dry Measure. Hence, a gallon is equal to about 4¼ Liters, and a bushel about 35 Liters. The Hektoliter (100 Liters) equals 2.8379 bushels.

The ***Unit of Weight*** is the ***Gram***. Its weight is 15.4324874 Troy grains, equal to the weight of a Cubic Centimeter of distilled water at a temperature at which ice melts. The denominations are shown in the following Table:

MEASURES OF WEIGHT.

1 Milligram (Mg.)		=	.0154324874 gr. Troy
10 Milligrams	= 1 Centigram (Cg.)	=	.154324874 "
10 Centigrams	= 1 Decigram (Dg.)	=	1.54324874 "
10 Decigrams	= 1 Gram (G.)	=	15.43234874 "
10 Grams	= 1 Dekagram (D.G.)	=	.3527398 oz. Avoir.
10 Dekagrams	= 1 Hektogram (H.G.)	=	3.527398 "
10 Hektograms	= 1 Kilogram (K.G.)	=	2.20462125 lbs. Avoir.
10 Kilograms	= 1 Myriagram (M.G.)	=	22.0462125 "
10 Myriagrams	= 1 Quintal (Q.)	=	220.462125 "
10 Quintals	= 1 Tonneau (T.)	=	2204.62125 "

For weighing light articles, the *Gram* is the standard unit. The weight of the five-cent coin adopted in 1866 is 5 grams. The *Kilogram* is the standard unit for weighing articles in common use, at which they are bought and sold. It is usually abbreviated. Thus, a *Kilo.* of salt or a *Kilo.* of meat means a *Kilogram* of salt and a *Kilogram* of meat. The *Quintal* and *Tonneau* are used for weighing very heavy articles.

The Equivalents of the principal Metric denominations in the various Units of the other weights and measures, and conversely, have been established by an Act of Congress. They are herewith presented in the following

TABLE OF EQUIVALENTS.

1 Centimeter = 0.3937 inch........1 Inch = 2.540 Centimeters.
1 Decimeter = 0.328 foot..........1 Foot = 3.048 Decimeters.
1 Meter = 1.0936 yds. (39.37 in.)...1 Yard = 0.9144 Meter.
1 Dekameter = 1.9884 rods ....1 Rod = 0.5029 Dekameter.
1 Kilometer = 0.62137 miles...... 1 Mile = 1.6093 Kilometers.
1 Sq. Centimeter = 0.1550 sq. inch.1 Sq. Inch = 6.452 Sq. Centimeters.
1 Sq. Decimeter = 0.1076 sq. foot..1 Sq. Foot = 9.2903 Sq. Decimeters.
1 Sq. Meter = 1.196 sq. yards ....1 Sq. Yard = 0.8361 Sq. Meter.
1 Are = 3.954 sq. rods............1 Sq. Rod = 25.293 Sq. Meters.
1 Hektare = 2.471 acres..........1 Acre = 0.4047 Hektare.
1 Sq. Kilometer = 0.3861 sq. mile..1 Sq. Mile = 2.590 Sq. Kilometers.
1 Cu. Centimeter = 0.0610 cu. inch.1 Cu. Inch = 16.387 Cu. Centimeters.
1 Cu. Decimeter = 0.0353 cu. foot,.1 Cu. Foot = 28.317 Cu. Decimeters
1 Cu. Meter = 1.308 cu. yards....1 Cu. Yard = 0.7645 Cu. Meter.
1 Stere = 0.2759 cord............1 Cord = 3.624 Steres.
1 Litre = 1.0567 liquid quart......1 Liquid Quart = 0.9463 Liter.
1 Dekaliter = 2.6417 gallons.......1 Gallon = 0.3785 Dekaliter.

METRIC SYSTEM. 303

1 Liter = 0.908 dry quart.........1 Dry Quart = 1.101 Liters.
1 Dekaliter = 1.135 pecks.........1 Peck = 0.881 Dekaliter.
1 Hektoliter = 2.8375 bushels......1 Bushel = 3.524 Hektoliters.
1 Gram = .03527 oz. Avoir........1 Ounce Avoir. = 28.35 Grams.
1 Kilogram = 2.2046 lb. Avoir.....1 Pound Avoir. = 0.4536 Kilogram.
1 Metric Ton = 1.1023 tons(2000 lbs).1 Ton (2000 lbs.)=0.9072 Metric Ton.
1 Gram = 15.432 gr. Troy.........1 Grain Troy = 0.0648 Gram.
1 Gram = 0.03215 oz. Troy.......1 Ounce Troy = 31.1035 Grams.
1 Kilogram = 2.679 lbs. Troy......1 Pound Troy = 0.3732 Kilogram.

The following denominations, though not strictly accurate, will be found convenient if committed to memory:

TABLE OF APPROXIMATE EQUIVALENTS.

1 Decimeter = 4 inches.
1 Meter = $1\tfrac{1}{10}$ yds. (3 ft. 3⅜ in.)
1 Dekameter = 2 rods.
1 Kilometer = ⅝ mile.
1 Are = 4 sq. rods ($\tfrac{1}{40}$ A).
1 Hektare = 2½ acres.

1 Stere = ¼ cord.
1 Liter = $1\tfrac{1}{18}$ liquid qt., $\tfrac{9}{10}$ dry qt.
1 Dekaliter = 1¼ pecks.
1 Hektoliter = $2\tfrac{7}{8}$ bushels.
1 Gram = 15½ grains.
1 Kilogram = 2⅕ pounds Av.

The *Numeration* and *Notation* of the Metric System being based on the Decimal Scale, the different Units, when expressed by figures, may be readily reduced to a common denomination, and added, subtracted, multiplied, or divided, as in Decimal Fractions.

Thus, 8 Meters, 7 Decimeters, 6 Centimeters, and 4 Millimeters, when expressed in Meters, would be written 8.764 M. If expressed in Centimeters, would be written 87.64 Cm., etc. Decimeters, Centimeters, and Millimeters bear the same relation to Meters as do Dimes, Cents, and Mills to Dollars.

EXERCISES IN NUMERATION AND NOTATION.

Express by Written Words.   *Express by Written Words.*

1. 7 M.   *4.* 6.82 Cm.   *7.* 45.04 Dm.   *10.* 3.68 Dm.
2. 4 Cm.   *5.* 1.45 M.   *8.* 3.462 Cm.   *11.* 48 M.
3. 14 Dm. *6.* 282 Mm.   *9.* 4.68 M.   *12.* .052 Mm.

Express by Figures and Abbreviations.

1. Twelve centigrams.
2. Twenty-four milligrams.
3. Three hundred and fourteen steres.
4. Eight hektares.
5. Two hundred hektoliters.
6. Five hektares.

To REDUCE Units of the Metric System to Equivalents of the Common System.

RULE.—*First find in the Table the Equivalent of one of the given* **Metric Units,** *multiply this Equivalent by the entire number of given* **Metric Units.** *If necessary, proceed as in Reduction of Compound Numbers.*

WRITTEN EXAMPLES.

OPERATION.

1 Liter = 1.0567 qts.
             36.5
           —————
           52835
           63402
           31701
        4)38.56955
        —————
        9.64238753

1. Reduce 36.5 liters to gallons.

Explanation.—We find in the Table of Equivalents that 1 Liter equals 1.0567 liquid quarts. Therefore, we multiply by 36.5, or the number of given Liters, and obtain 38.56955 quarts. Since the number of gallons is required, we divide by 4, and obtain 9.64238753 gallons.

Reduce

2. 8 meters to yards.
3. 5 kilograms to pounds Avoir.

Reduce

4. 24 sq. meters to sq. yards.
5. 9 steres to cords.

To REDUCE Units of the Common System to Equivalents of the Metric System.

RULE.—*First find in the Table the Equivalent of one of the given Units of the* **Common System.** *Multiply this Equivalent by the entire number of* **Common Units.** *If necessary, proceed as in Reduction of Compound Numbers.*

METRIC SYSTEM.

WRITTEN EXAMPLES.

OPERATION.
1 bushel = 3.524 dek.
12
———
42.288

1. Reduce 12 bushels to dekaliters.

Explanation.—We find in the Table of Equivalents that 1 Bushel equals 3.524 Dekaliters. Therefore, we multiply by 12, and obtain 42.288 Dekaliters, or the required number of Metric Units.

Reduce
2. 16 cords to steres.
3. 20 sq. inches to sq. centimeters.

Reduce
4. 6 kilograms to pounds Avoir.
5. 9 yards to meters.

To find the CONVERSE DECIMAL Equivalent when the Decimal Equivalent of a Unit of either the Common or Metric System is given.

RULE.—*Divide 1 by the given* **Decimal Equivalent** *and the Quotient will be the* **Converse Decimal** *required.*

WRITTEN EXAMPLES.

OPERATION.
.3937)10000(2.54
7874
———
21260
19685
———
15750
15748

1. If a centimeter is equal to .3937 of an inch, how many centimeters are equal to an inch?

Explanation.—Since the Converse Decimal Equivalent is the Decimal of the Reciprocal of the given Equivalent, we divide 1 by the given Equivalent as in the operation, and obtain 2.54 Centimeters, or the number which is equal to 1 inch.

2. If 1 grain Troy equals 0.648 of a gram, how many grains are equal to a gram?

3. A peck equals .881 of a dekaliter. How many pecks are equal to a dekaliter?

4. A gram equals .03215 of an ounce Troy. How many grams are equal to an ounce?

5. A stere equals .2759 of a cord. How many steres are equal to a cord?

TRADE DISCOUNTS

Trade Discounts are allowances, or abatements, made by manufacturers and jobbers from their marked or list prices.

It is customary in some branches of business, for merchants or manufacturers to offer their goods or wares at list prices, subject to abatement by certain series of discounts.

TRADE DISCOUNTS. 307

In the marginal illustration, the bill, or list price, amounts to $240, less the series 20% — 20% — 20% — 10% — 10% — 5%, and shows a net cost of $94.55, which, deducted from the list price, leaves a discount of $145.45.

To those not familiar with the principles governing Trade Discounts, it might be supposed that the rates of discount on the above invoice equal 85%, but such is not the case. It will be seen from inspection, that the first rate of discount *only* is deducted from the *List Price*, and that the subsequent rates are deducted from the differences in the order shown thereon, leaving the real discount of $145.45, but $60\frac{29}{100}\%$.

WRITTEN EXAMPLES.

1. A dealer bought a quantity of hardware, at $480, at a discount of 20, 10 and 5¢. What was the net cost?

Operation.

Invoice price	$480
Less 20¢ or ⅕	96
	$384
Less 10¢ or ¹⁄₁₀	38.40
	$345.60
Less 5¢ or ¹⁄₂₀	17.28
Net,	$323.32

Find the Net Cost.

	Invoice Price.	*Discount off.*
2.	$460.40.	20, 10 & 10%.
3.	$300.00.	40, 5 & 6%.
4.	$860.78.	30, 10 & 4%.
5.	$900.00.	20, 5, 2½%.
6.	$682.34.	20, 10, & 5%.
7.	$250.41.	30 & 10%.
8.	$340.80.	45 & 5%.
9.	$200.00.	40, 10, & 5%.

Note.—In deducting the discounts, when the fraction of a cent less than one-half remains, it is usually disregarded, and the whole cent is taken when equal to one-half.

The Clearing House

[*From Sadler's Counting House Arithmetic.*]

MAKING THE EXCHANGE AT CLEARING HOUSE.

Is an institution established in the larger cities of the country, where are assembled on each banking day the representatives of the various Banks, members of the Clearing House Association, for the purpose of effecting exchanges and the settlement of balances between each other.

In smaller cities the business of settlement is made direct.

A ***Bank Clearing House Association*** is an association of Banks organized for convenience in making daily settlements of their Bank Accounts, and for the purpose of promoting their general welfare.

The first Clearing House was established in the City of New York, on October 1st, 1853. Previous to that time there were about sixty or more Banks in operation, and in consequence of the inconvenience, annoyance, and great loss of time in making daily settlements, it became a general agreement that settlements should be made only on every Friday. At that time there were no Coin Certificates, and hundreds of pounds of silver were brought to the place of settlement weekly.

There appeared in 1841 a pamphlet, published by Albert Gallatin, entitled "Suggestions on the Banks and Currency of the Several United States, in reference principally to the Suspension of Specie Payments," from which we insert a few paragraphs of the same bearing directly upon the necessity of a Clearing House.

"There is a measure which, though belonging to the administration of Banks, rather than to legal enactments, is suggested on account of its great importance. Few regulations would be more useful in preventing dangerous expansions of discounts and issues on the part of the City Banks, than a regular exchange of Notes and Checks, and an actual daily or semi-weekly payment of the balances. It must be recollected, that it is by this process alone that a bank of the United States has ever acted, or been supposed to act, as a regulator of the currency. Its action would not, in that respect, be wanted in any city, the banks of which would, by adopting the process, regulate themselves. It is one of the principal ingredients of the system of the banks of Scotland. The bankers of London, by the daily exchange of Drafts at the Clearing House, reduce the ultimate balance to a very small sum; and that balance is immediately paid in notes of the Bank of England. The want of a similar arrangement among the banks of this city produces relaxation, favors improper expansions, and is attended with serious inconveniences. The principal difficulty in the way of an arrangement for that purpose, is the want of a common medium other than specie for effecting the payment of balances. These are daily fluctuating; and a perpetual drawing and redrawing of specie from and into the banks is unpopular and inconvenient."

"In order to remedy this, it as been suggested that a general *Cash Office* might be established, in which each Bank should place a sum in specie, proportionate to its capital, which would be carried to its credit in the books of the office. Each Bank would be daily debited or credited in those books for the balance of its account with all the other Banks. Each Bank might at any time draw for specie on the office for the excess of its credit beyond its quota, and each Bank should be obliged to replenish its quota whenever it was diminished one-half, or in any other proportion agreed on."

Although the New York Clearing House plan went into effect in October, 1853, yet no constitution was adopted until June 6, 1854. This first constitution, and a subsequent amendment to the same, were prepared by George Curtis, Esq. The following Extracts from the Constitution of the Clearing House Association of Philadelphia will be found to embrace the essential features suggested in the pamphlet before alluded to, with such modifications as the changes in the currency, banking laws, etc., demand.

"The title of this Association shall be the Clearing House Association of Philadelphia. Its object shall be to effect at one place the Daily Exchanges and the Runners' Exchanges between the several Associated Banks, and the payment at the same place, of the balances resulting from such exchanges. But the Association shall be in no wise responsible in regard to the balances resulting therefrom, except so far as such balances shall be actually paid into the hands of the manager. The responsibility of the Association is strictly limited to the faithful distribution by the Manager, among the Creditor Banks for the time being, of the sums actually received by him; and should any losses occur while the said balances are in the custody of the Manager, they shall be borne and paid by the Associated Banks in the same proportion as the other expenses of the Clearing House, as hereinafter provided for."

"The Associated Banks shall, from time to time, appoint one of their number to be a depository to receive in special trust such Coin or United States legal tender Notes as any of the Associated Banks may choose to send to it for safe-keeping. The depository shall issue certificates in exchange for such Coin or United States legal tender Notes, in proper form and for convenient amounts. Such certificates shall be negotiated only among the Associated Banks, and shall be received by them in payments of balances at the Clearing House. Such special deposits of Coin or United States legal tender Notes are to be entirely voluntary, each Bank being left perfectly free to make them or not, at its own discretion.

"The Coin or United States legal tender Notes, thus placed in special deposit, is to be the absolute property of such of the Associated Banks as shall from time to time be the holders of the certificates, and is to be held by the depositors, subject to withdrawal on the presentation of the proper certificates, at any time during banking hours."

The daily routine in the several Clearing Houses is about as follows: each Bank of the Association sends to the Clearing House both a Messenger and a Clerk, the former carrying with him such Checks, Drafts, Notes, etc., as the Bank he represents may have against the Banks in the Association, and the latter bringing a formal "Clerk's Statement" (see Form 1), also a Credit Ticket (see Form 2), showing the amount due from all other Banks; and on presentation of the Ticket to the Manager, he is credited on the Clearing House Proof (see Form 3).

At the appointed time the Clerks and Messengers take the places assigned them as the representatives of their respective Banks, the Messenger on the outside and the Settling Clerk on the inside of an oblong desk, numbered consecutively, and large enough to accommodate the entire number of Banks in the Association. At the direction of the Manager, the Messenger at Desk No. 1 passes to the Settling Clerk, at, say Desk No. 28—if twenty-eight Banks be represented—and hands over to him a Clerk's Statement containing Debit Amounts of each Bank, and also surrenders the vouchers which he holds against the Bank represented by the Settling Clerk. The Clerk then verifies the figures by comparing those of the package with the state-

[FORM 1.]
PHILADELPHIA CLEARING HOUSE,
FROM
FIRST NATIONAL BANK.
Clerk's Statement, June 1, 188 .

No.	Names of Banks.	Dr. Am'ts.	Rec's.
1	PHILADELPHIA NATIONAL,	Amount of Checks brought by Messenger from the First National Bank. Aggregate amount to go to their Credit at Clearing House. Similar Statements are sent by all Bank members of the Clearing House Association.	In this space, the Receipt of Checks is acknowledged by the Clerks from the different Banks.
2	NORTH AMERICA,		
3	FARMERS' AND MECHANICS' NATIONAL,		
4	COMMERCIAL NATIONAL,		
5	MECHANICS' NATIONAL,		
6	NATIONAL BANK N. LIBERTIES,		
7	SOUTHWARK NATIONAL,		
8	KENSINGTON NATIONAL,		
9	PENN NATIONAL,		
10	WESTERN NATIONAL,		
11	MANUFACTURERS' NATIONAL,		
12	NATIONAL BANK OF COMMERCE,		
13	GIRARD NATIONAL,		
14	TRADESMEN'S NATIONAL,		
15	CONSOLIDATION NATIONAL,		
16	CITY NATIONAL,		
17	COMMONWEALTH NATIONAL,		
18	CORN EXCHANGE NATIONAL,		
19	UNION NATIONAL,		
20	THIRD NATIONAL,		
21	SIXTH NATIONAL,		
22	SEVENTH NATIONAL		
23	EIGHTH NATIONAL,		
24	CENTRAL NATIONAL,		
25	NATIONAL BANK OF THE REPUBLIC,		
26	NATIONAL SECURITY,		
27	CENTENNIAL NATIONAL,		
28	FIDELITY NATIONAL,		
	Total Credit,	$	

ment; and if correct, places the amount in the column under Creditor Banks in the following *Settling Clerk's Statement* (Form 3), and signs his name opposite that of his Bank on the "*Clerk's Statement*" (Form 1), and returns it to the Messenger, who then passes to No. 27. In like manner each Messenger and Clerk proceed until all the Messengers arrive at their respective desks, when they hand over to their own Clerks

[*FORM 2.*]
PHILADELPHIA CLEARING HOUSE.

```
No. ................                June 1, 188 .

        Credit FIRST NATIONAL BANK.

Amount of Checks per Messenger, - - - $

                        ..................................Teller.
```

the statements signed by the representatives of the several Banks, leaving the Settling Clerks to finish the business of their Banks. The Settling Clerks then find the difference between the total of "Banks Cr." and that of "Banks Dr.," and then make out a "balance ticket."

If the balance is in favor of the other Banks, the "Debit Balance due Clearing House" is written thereon. If the balance is due from other banks represented, then the words "Credit Balance due" is extended. The amounts due the Clearing House must be next paid in, after which the Banks having balances due them are paid

THE CLEARING HOUSE.

[FORM 3.]

PHILADELPHIA CLEARING HOUSE.
FIRST NATIONAL BANK SETTLEMENT.

No. 1.] Settling Clerk's Statement, June 1, 188 . [8½ A. M.

Dr. Balance.	Banks Debtor.	No.	BANKS.	Total Bank Cr.	Credit Balance.
$83,376.27	$53,965.27		In this column appear the balances due from other Banks. / In this column appear the DEBITS against the various Banks, showing the aggregate at the footing.		
		1	FIRST NATIONAL BANK,		
		2	NORTH AMERICA,		
		3	FARM'S & MECH'S NATIONAL,		
		4	COMMERCIAL NATIONAL,		
		5	MECHANICS' NATIONAL,		
		6	NATIONAL BK. N. LIBERTIES,		
		7	SOUTHWARK NATIONAL,		
		8	KENSINGTON NATIONAL,		
		9	PENN NATIONAL,		
		10	WESTERN NATIONAL,		
		11	MANUFACTURERS' NATIONAL,		
		12	NATIONAL BK. OF COMMERCE,		
		13	GIRARD NATIONAL,		
		14	TRADESMEN'S NATIONAL,		
		15	CONSOLIDATION NATIONAL,		
		16	CITY NATIONAL,		
		17	COMMONWEALTH NATIONAL,		
		18	CORN EXCHANGE NATIONAL,		
		19	UNION NATIONAL,		
		20	THIRD NATIONAL,		
		21	SIXTH NATIONAL,		
		22	SEVENTH NATIONAL,		
		23	EIGHTH NATIONAL,		
		24	CENTRAL NATIONAL,		
		25	NAT. BANK OF THE REPUBLIC,		
		26	NATIONAL SECURITY,		
		27	CENTENNIAL NATIONAL,		
		28	FIDELITY NATIONAL,		
			BALANCE TO CR.	39,634.19	9,237.64
			AGGREGATE,	14,134.08	14,134.08
			In this column appear the CREDITS in favor of each of the various Banks, showing the aggregate at the footing.	$53,965.27	$23,376.27
			In this column appear the balance due to other Banks.		

therefrom, within a certain hour on the same day.

The time consumed in settling the daily balances is usually less than *ten minutes*. In the City of New York there is daily entered on the

```
[FORM 4.]
PHILADELPHIA CLEARING HOUSE,
No.............                    June 1, 188  .
        Amount Received,    -   -  $53965.27
        Amount Returned,    -   -   39834.19
        Balance to Credit—(name of bank), -  ——— $14131.08
        Balance to Debit—(   "    "   "  ), -
                      ..................Settling Clerk.
```
(BALANCE TICKET.)

Clearing House Proof upwards of one hundred millions of dollars, but the actual balance paid over to the respective Banks is often less than two millions.

The following is a form of a Clearing House Proof, showing an aggregate of more than seven million dollars, which has been settled through the Clearing House by the payment of less than one hundred thousand dollars. The accuracy of the Proof is determined by the footings of the Columns "Debtor Bank" and "Total Debtor" agreeing with those of the "Creditor Banks" and "Total Credits."

The advantages derived from the use of the Clearing House are briefly stated in an appendix to Cleveland's Banking Laws of New York, as follows:

"On the day when the Clearing House began business about twenty-seven hundred open active accounts on the

PHILADELPHIA CLEARING HOUSE.

Clearing House Proof.   June 1, 188 .

No.	Debtor Bank.	Total Debits.	BANKS.	No.	Total Crs.	Cr. Bank.
	In this column appear all the balances due the Clearing House from the Debtor Banks. The aggregate sums appear at the footing.	In this column appear the TOTAL DEBITS against the various Banks, showing the aggregate at the footing.	FIRST NATIONAL BANK, NORTH AMERICA, FARM'S & MECH'S NATIONAL, COMMERCIAL NATIONAL, MECHANICS' NATIONAL, NATIONAL BK. N. LIBERTIES, KENSINGTON NATIONAL, PENN NATIONAL, MANUFACTURERS' NATIONAL, NATIONAL BK. OF COMMERCE, TRADESMEN'S NATIONAL, CITY NATIONAL, COMMONWEALTH NATIONAL, CORN EXCHANGE NATIONAL, UNION NATIONAL, THIRD NATIONAL, SIXTH NATIONAL, SEVENTH NATIONAL,	1 2 3 4 5 6 7 8 9 10 11 12 13 14 15 16 17 18	In this column appear the TOTAL CREDITS in favor of the various Banks, showing the aggregate at the footing.	In this column appear the balances due the Credit Banks by the Clearing House. The aggregate of sums appear at the footing.
	$98,463.71	$7,465,237.54	AGGREGATE,		$7,465,237.54	$98,463.71

ledgers of the Associated Banks were balanced—the most of them for the first time, and all of them finally. The business which had rendered necessary this large number of accounts was thenceforth accomplished more quickly, with less annoyance to Bank officers, and with greater safety to all concerned. The results may be briefly enumerated as follows:

"*First*—The condensation for each Bank of forty-eight balances into one, and the settlement of that balance without a movement of specie.

"*Secondly*—The avoidance of numerous accounts, entries, and postings.

"*Thirdly*—Great saving of time to the Porters, and of risk in making exchanges and settlements from bank to bank.

"*Fourthly*—Relief from a vast amount of labor and annoyance to which the great army of Cashiers, Tellers, and Book-keepers were subjected under the old system.

"*Fifthly*—The liberation of the Associated Banks from all injurious dependence on each other.

"*Sixthly*—The absolute facility afforded by the books of the Clearing House for knowing at all times the management and standing of every Bank in the Association."

WAGES—VALUE OF TIME.

For DAYS *and* HOURS, *at Stated Rates Per Week.*

Rate	$3	3½	$4	4½	$5	5½	$6	6½	$7	7½	$8	$9	10	11	12
Hours 1	5	6	7	8	8	9	.10	.11	.12	.13	.13	.15	.17	.18	.20
2	.10	.12	.13	.15	.17	.18	.20	.22	.23	.25	.27	.30	.33	.37	.40
3	.15	.18	.20	.23	.25	.28	.30	.33	.35	.38	.40	.45	.50	.55	.60
4	.20	.23	.27	.30	.33	.37	.40	.43	.47	.50	.53	.60	.67	.73	.80
5	.25	.29	.33	.38	.42	.46	.50	.54	.58	.63	.67	.75	.83	.92	1.00
6	.30	.35	.40	.45	.50	.55	.60	.65	.70	.75	.80	.90	1.00	1.10	1.20
7	.35	.41	.47	.53	.58	.64	.70	.76	.82	.88	.93	1.05	1.17	1.28	1.40
8	.40	.47	.53	.60	.67	.73	.80	.87	.93	1.00	1.07	1.20	1.33	1.47	1.60
9	.45	.53	.60	.68	.75	.83	.90	.98	1.05	1.13	1.20	1.35	1.50	1.65	1.80
Days 1	.50	.58	.67	.75	.83	.92	1.00	1 08	1.17	1.25	1.33	1.50	1.67	1.83	2.00
2	1.00	1.17	1.33	1.50	1.67	1.83	2.00	2.17	2.33	2.50	2.67	3.00	3.33	3.67	4.00
3	1.50	1.75	2.00	2.25	2.50	2.75	3.00	3.25	3.50	3.75	4.00	4.50	5.00	5.50	6.00
4	2.00	2.33	2.67	3.00	3.33	3.67	4.00	4.33	4.67	5.00	5.33	6.00	6.67	7.33	8.00
5	2.50	2.92	3.33	3.75	4.17	4.58	5.00	5.42	5.83	6.25	6.67	7.50	8.33	9.17	$10.

For DAYS, *at Stated Rates Per Month.*

Rate.	$14	$15	$16	$17	$18	$19	$20	$21	$22	$23	$24	$25
Days 1	.54	.58	.62	.65	.69	.73	.77	.81	.85	.88	.92	.96
2	1.08	1.15	1.23	1.31	1.38	1.46	1.54	1.62	1.69	1.77	1.85	1.92
3	1.62	1.73	1.85	1.96	2.08	2.19	2.31	2.42	2.54	2.65	2.77	2.88
4	2.15	2.31	2.46	2.62	2.77	2.92	3.08	3.23	3.38	3.54	3.69	3.85
5	2.69	2.88	3.08	3.27	3.46	3.65	3.85	4.04	4.23	4.42	4.62	4.81
6	3.23	3.46	3.69	3.92	4.15	4.38	4.62	4.85	5.08	5.31	5.54	5.77
7	3.77	4.04	4.31	4.58	4.85	5.12	5.38	5.65	5.92	6.19	6.46	6.73
8	4.31	4.62	4.92	5.23	5.54	5.85	6.15	6.46	6.77	7.08	7.38	7.69
9	4.85	5.19	5.54	5.88	6.23	6.58	6.92	7.27	7.62	7.96	8.31	8.65
10	5.38	5.77	6.15	6.54	6.92	7.31	7.69	8.08	8.46	8.85	9.23	9.62
11	5.92	6.35	6.77	7.19	7.62	8.04	8.46	8.88	9.31	9.73	10.15	10.58
12	6.46	6.92	7.38	7.85	8.31	8.77	9.23	9.69	10.15	10.62	11.08	11.54
13	7.00	7.50	8.00	8.50	9.00	9.50	10.00	10.50	11.00	11.50	12.00	12.50
14	7.54	8.08	8.62	9.15	9.69	10.23	10.77	11.31	11.85	12.38	12.92	13.46
15	8.08	8.65	9.23	9.81	10.38	10.96	11.54	12.12	12.69	13.27	13.85	14.42
16	8.62	9.23	9.85	10.46	11.08	11.69	12.31	12.92	13.54	14.15	14.77	15.38
17	9.15	9.81	10.46	11.12	11.77	12.42	13.08	13.73	14.38	15.04	15.69	16.35
18	9.69	10.38	11.08	11.77	12.46	13.15	13.85	14.54	15.23	15.92	16.62	17.31
19	10.23	10.96	11.69	12.42	13.15	13.88	14.62	15.35	16.08	16.81	17.54	18.27
20	10.77	11.54	12.31	13.08	13.85	14.62	15.38	16.15	16.92	17.69	18.46	19.23
21	11.31	12.12	12.92	13.73	14.54	15.35	16.15	16.96	17.77	18.58	19.38	20.19
22	11.85	12.69	13.54	14.38	15.23	16.08	16.92	17.77	18.62	19.46	20.31	21.15
23	12.38	13.27	14.15	15.04	15.92	16.81	17.69	18.58	19.46	20.35	21.23	22.12
24	12.92	13.85	14.77	15.69	16.62	17.54	18.46	19.38	20.31	21.23	22.15	23.08
25	13.46	14.42	15.38	16.35	17.31	18.27	19.23	20.19	21.15	22.12	23.08	24.04
26	14.00	15.00	16.00	17.00	18.00	19.00	20.00	21.00	22.00	23.00	24.00	25.00

HOW TO OBTAIN WEALTH.

Table showing the net amount of earnings of One Cent to Twenty-five Dollars per Day for Ten Years of 313 working days, without interest, and with interest at 6, 7, and 8 per cent., compounded each Six Months.

Savings per day	Without interest.	With interest at 6 per cent.	With interest at 7 per cent.	With interest at 8 per cent.
1	$31 13	$42 05	$44 26	$46 60
2	62 26	84 10	88 52	93 21
3	93 39	126 16	132 77	139 81
4	124 52	168 21	177 03	186 41
5	156 50	210 26	222 29	233 01
6	187 80	252 31	265 56	279 62
7	219 10	294 36	309 80	326 22
8	250 40	336 52	354 06	372 82
9	281 70	378 47	398 22	419 43
10	313 00	420 52	442 58	466 03
15	469 50	630 78	666 87	699 04
20	626 00	841 04	885 15	932 05
25	782 50	1,051 30	1,111 44	1,165 07
30	939 00	1,261 56	1,327 73	1,398 08
40	1,252 00	1,682 09	1,770 31	1,864 11
50	1,565 00	2,102 61	2,212 89	2,330 13
60	1,878 00	2,523 13	2,655 46	2,796 16
70	2,191 00	2,943 65	3,098 04	3,262 19
80	2,504 00	3,364 17	3,540 62	3,728 22
90	2,817 00	3,784 69	3,982 19	4,194 24
$1 00	3,130 00	4,205 21	4,425 77	4,660 27
2 00	6,260 00	8,410 43	8,851 54	9,320 54
3 00	9,390 00	12,615 64	13,277 31	13,980 81
4 00	12,520 00	16,820 85	17,703 08	18,641 08
5 00	15,650 00	21,026 07	22,228 85	23,301 35
6 00	18,750 00	25,251 28	26,554 32	27,961 62
7 00	21,910 00	29,436 50	30,980 39	32,621 89
8 00	25,040 00	33,642 71	35,406 16	37,282 14
9 00	28,170 00	37,846 92	39,821 93	41,942 42
10 00	31,300 00	42,052 14	44,257 70	46,602 69
15 00	46,950 00	63,078 20	66,668 55	69,904 04
20 00	62,600 00	84,104 27	88,515 40	93,205 39
25 00	78,250 00	105,030 00	111,144 00	116,507 60

From the above Table it can readily be observed why "Fortunes are Spent by Trifles," and the advantage in saving, if one desires to obtain a competency. This Table is worthy the careful attention of our young men who desire success in life.

VALUABLE TABLES

For the Merchant, Farmer, and Purchaser, showing at sight the Value of Articles Sold by the Pound, Dozen, Yard, or Piece, as Groceries, Produce, Dry Goods, etc.

These Tables embody nearly all of the practical features comprised in publications devoted exclusively to the subject of Ready Reckoning, for which prices are asked nearly equal to the cost of this entire work. They will be found invaluable in ascertaining the value of articles usually sold by the Business Trader and Farmer and consumed in families.

APPLICATION OF THE TABLES.

The outside perpendicular columns to the left and right show the price of the article, and the upper and lower lines the *quantity*.

When advisable in securing results the working can be reversed.

EXAMPLE.—What will 14 lbs. of Coffee cost at 29 cts. per pound? See price in column to the left, 29, follow the finger along the line until the sum under the column 14 is reached, and you have the amount, $4.06.

NOTE I.—When the price or quantity required is not shown in the tables, reduce the number of either or both to such amounts as are shown in the extremes; ascertain the product from the table, and multiply by the number or numbers used as a divisor in reduction of the original amounts.

EXAMPLE.—What will 450 bushels of screenings cost at 23 cts. per bushel? As 450 is not contained in the table, we reduce it to 45 by dividing by 10. Referring to 45 in column at the left, and on the same line under the head number 23, we have 10.35, the cost of 45 bushels. $10.35 \times 10 = \$103.50$, the cost of 450 bushels.

NOTE II.—When the price or quantity is a fractional part of a whole number, ascertain the amount from the table by using the numerator as a whole number, and then divide by the denominator, adding the result to the product already obtained.

TABLE OF READY CALCULATIONS.

See page 321

1	2	3	4	5	6	7	8	9	10	11	12	13	14	15
12	24	36	48	60	72	84	96	108	120	132	144	156	168	180
13	26	39	52	65	78	91	104	117	130	143	156	169	182	195
14	28	42	56	70	84	98	112	126	140	154	168	182	196	210
15	30	45	60	75	90	105	120	135	150	165	180	195	210	225
16	32	48	64	80	96	112	128	144	160	176	192	208	224	240
17	34	51	68	85	102	119	136	153	170	187	204	221	238	255
18	36	54	72	90	108	126	144	162	180	198	216	234	252	270
19	38	57	76	95	114	133	152	171	190	209	228	247	266	285
20	40	60	80	100	120	140	160	180	200	220	240	260	280	300
21	42	63	84	105	126	147	168	189	210	231	252	273	294	315
22	44	66	88	110	132	154	176	198	220	242	264	286	308	330
23	46	69	92	115	138	161	184	207	230	253	276	299	322	345
24	48	72	96	120	144	168	192	216	240	264	288	312	336	360
25	50	75	100	125	150	175	200	225	250	275	300	325	350	375
26	52	78	104	130	156	182	208	234	260	286	312	338	364	390
27	54	81	108	135	162	189	216	243	270	297	324	351	378	405
28	56	84	112	140	168	196	224	252	280	308	336	364	392	420
29	58	87	116	145	174	203	232	261	290	319	348	377	406	435
30	60	90	120	150	180	210	240	270	300	330	360	390	420	450
31	62	93	124	155	186	217	248	279	310	341	372	403	434	465
32	64	96	128	160	192	224	256	288	320	352	384	416	448	480
33	66	99	132	165	198	231	264	297	330	363	396	429	462	495
34	68	102	136	170	204	238	272	306	340	374	408	442	476	510
35	70	105	140	175	210	245	280	315	350	385	420	455	490	525
36	72	108	144	180	216	252	288	324	360	396	432	468	504	540
37	74	111	148	185	222	259	296	333	370	407	444	481	518	555
38	76	114	152	190	228	266	304	342	380	418	456	494	532	570
39	78	117	156	195	234	273	312	351	390	429	468	507	546	585
40	80	120	160	200	240	280	320	360	400	440	480	520	560	600
41	82	123	164	205	246	287	328	369	410	451	492	533	574	615
42	84	126	168	210	252	294	336	378	420	462	504	546	588	630
43	86	129	172	215	258	301	344	387	430	473	516	559	602	645
44	88	132	176	220	264	308	352	396	440	484	528	572	616	660
45	90	135	180	225	270	315	360	405	450	495	540	585	630	675
46	92	138	184	230	276	322	368	414	460	506	552	598	644	690
47	94	141	188	235	282	329	376	423	470	517	564	611	658	705
48	96	144	192	240	288	336	384	432	480	528	576	624	672	720
49	98	147	196	245	294	343	392	441	490	539	588	637	686	735
50	100	150	200	250	300	350	400	450	500	550	600	650	700	750
1	2	3	4	5	6	7	8	9	10	11	12	13	14	15

TABLE OF READY CALCULATIONS.

See page 321

16	17	18	19	20	21	22	23	24	25	
192	204	216	228	240	252	264	276	288	300	12
208	221	234	247	260	273	286	299	312	325	13
224	238	252	266	280	294	308	322	336	350	14
240	255	270	285	300	315	330	345	360	375	15
256	272	288	304	320	336	352	368	384	400	16
272	289	306	323	340	357	374	391	408	425	17
288	306	324	342	360	378	396	414	432	450	18
304	323	342	361	380	399	418	437	456	475	19
320	340	360	380	400	420	440	460	480	500	20
336	357	378	399	420	441	462	483	504	525	21
352	374	396	418	440	462	484	506	528	550	22
368	391	414	437	460	483	506	529	552	575	23
384	408	432	456	480	504	528	552	576	600	24
400	425	450	475	500	525	550	575	600	625	25
416	442	468	494	520	546	572	598	624	650	26
432	459	486	513	540	567	594	621	648	675	27
448	476	504	532	560	588	616	644	672	700	28
464	493	522	551	580	609	638	667	696	725	29
480	510	540	570	600	630	660	690	720	750	30
496	527	558	589	620	651	682	713	744	775	31
512	544	576	608	640	672	704	736	768	800	32
528	561	594	627	660	693	726	759	792	825	33
544	578	612	646	680	714	748	782	816	850	34
560	595	630	665	700	735	770	805	840	875	35
576	612	648	684	720	756	792	828	864	900	36
592	629	666	703	740	777	814	851	888	925	37
608	646	684	722	760	798	836	874	912	950	38
624	663	702	741	780	819	858	897	936	975	39
640	680	720	760	800	840	880	920	960	1000	40
656	697	738	779	820	861	902	943	984	1025	41
672	714	756	798	840	882	924	966	1008	1050	42
688	731	774	817	860	903	946	989	1032	1075	43
704	748	792	836	880	924	968	1012	1056	1100	44
720	765	810	855	900	945	990	1035	1080	1125	45
736	782	828	874	920	966	1012	1058	1104	1150	46
752	799	846	893	940	987	1034	1081	1128	1175	47
768	816	864	912	960	1008	1056	1104	1152	1200	48
784	833	882	931	980	1029	1078	1127	1176	1225	49
800	850	900	950	1000	1050	1100	1150	1200	1250	50
16	17	18	19	20	21	22	23	24	25	

SADLER'S ✧ ARITHMETICS
ARE THE BEST.

SADLER'S COUNTING-HOUSE ARITHMETIC.

∴ The standard work of reference for the Counting-Room, and more extensively used in Business Colleges than any similar publication. ∴ ∴ ∴ ∴ ∴ ∴ ∴ ∴ ∴ ∴ ∴ ∴

SADLER'S COMMERCIAL ARITHMETIC.

∴ For Business Colleges and Commercial Departments of Literary Colleges and High Schools, is unsurpassed. Every teacher of Business Arithmetic will be delighted with this volume, for it contains just what he needs; no more and no less.

SADLER'S COMMERCIAL ARITHMETIC.

∴ The School Edition of this work is prepared exclusively for Colleges, High Schools and Academies. ∴ ∴ ∴ ∴ ∴

SADLER'S INDUCTIVE ARITHMETIC.

∴ For Business Universities, Normal and High Schools. This is the largest and most complete work on Arithmetical subjects ever published, and should be in the hands of every *live* teacher as a desk copy. ∴ ∴ ∴ ∴ ∴ ∴ ∴ ∴ ∴ ∴ ∴ ∴

SADLER'S HAND-BOOK OF ARITHMETIC.

∴ A book of over 4,000 practical Arithmetical Problems.

Copies of either of the above works will be sent to teachers for examination, post-paid, on receipt of wholesale price.
Correspondence and orders solicited.
For descriptive circulars and price-list, address:

W. H. SADLER,
10 AND 12 N. CHARLES STREET.   **BALTIMORE, MD.**

www.ingramcontent.com/pod-product-compliance
Lightning Source LLC
Chambersburg PA
CBHW030741230426
43667CB00007B/802